BUFF TEA

BUFF TEA

Edward M. Erdelac

Texas Review Press
Huntsville, Texas

Copyright © 2011 by Edward M. Erdelac
All rights reserved
Printed in the United States of America

FIRST EDITION, 2011
Requests for permission to reproduce material from this work should be sent to:

>   Permissions
>   Texas Review Press
>   English Department
>   Sam Houston State University
>   Huntsville, TX 77341-2146

Cover design by Virginia Houk

Front cover photo by Cody Sechelski

Author photo courtesy of Edward M. Erdelac

Library of Congress Cataloging-in-Publication Data

Erdelac, Edward M. (Edward Michael), 1975-
  Buff tea / Edward M. Erdelac. -- 1st ed.
    p. cm.
  ISBN 978-1-933896-62-5 (pbk. : alk. paper)
  1. West (U.S.)--Fiction. 2. American bison hunting--Fiction. 3. Frontier and pioneer life--West (U.S.)--Fiction. I. Title.
  PS3605.R34B84 2011
  813'.6--dc22
                            2010052951

*To God and my family,
and to Ty Cashion, whose dedication and
encouragement are perpetual*

BUFF TEA

PREAMBLE

The notion that taking a man's life to save one's soul from damnation might seem blasphemous, and probably not unlike the rationalizations of a guiltless psychopath. Nevertheless, once I killed a man for that very reason—not in the high fury of a sanctioned battle for God and democracy, but in the quiet stillness of deliberate murder. I guess murder was the farthest thing from my mind that long ago spring of eighteen seventy four when I was nineteen and yet still whole.
    I was born in Chicago. My father founded what turned out be a lucrative brick and cobble manufacturing company. While not overly wealthy, we were relatively unacquainted with want. As the son and daughter of immigrants, my parents were *nouveau riche,* and tried their hardest to infiltrate the sacrosanct mysteries of the upper class.
    My mother decided early on that I was to command the vanguard in her own personal assault on high society. I was groomed like a thoroughbred and goaded into a blur of social situations where I was expected to sit prim and strapping, flirt with the old ladies, and feign a gentlemanly interest in their tittering, fool daughters.
    When I came into my chin whiskers however, I saw that if I ever wanted to amount to anything of worth (other than the next generation purveyor of bricks and cobbles) I had best "leave behind the goose down pillows and Prince Albert coats of my youth and seek out such crucibles as had tempered the prophets to their predestined undertakings"

(that, a bit of forty dollar prose from my journal of the time). Men like Herman Melville and Henry Thoreau had inspired me to pursue my fortune with a pen and paper. I was to be a writer, come hell or high water.

Maybe I thought of myself as an aspiring prophet back then. I'm an old man now, and have moved no masses. But back then, brimming over with piss and vinegar, I was sure I was meant for "some great purpose that the soft fleshed geese of my generation could never be privy to" (more high handed scribblings from my riper days).

I went west because in my mind there could be no other course for a young man with visions so lofty as mine. The mythology I read in the Police Gazette and other periodicals of its sort had a hold of me. In my milk-fed mind the West was a fairy tale land where the blood of heroes mixed in the dusty streets of far-flung frontier towns, pouring from the veins of errant champions with lively monikers like 'Indian Dick,' 'Wild Bill Hickock,' and 'The Waco Kid.' I devoured racks of those cheap penny dreadfuls like most youngsters did jars of sweets at the corner drug store. Their crudely set type smeared in my eager hands like the hidden tracks of wily Indian braves. I fully expected to return from my travels a grizzled plainsman in a fur trimmed suit of buckskin with all the experiences I needed to foster my career as an author of wide renown.

Nearly sixty years have passed since I set out to begin my life's work, and this is the first I've written of the bloody deed that drove me from those palatial dreams of youth. I look back on that old journal as I write this now, and I hardly recognize the young man whose purple words line its pages. I left his aspirations behind as surely as I left the comfort of my life in the Midwest, as surely as I left a part of myself lying on a blackened patch of the Texas plains.

I set pen to paper now to relate the story of the strange salvation that came to me with the shedding of blood, and of the real mentors of my life; Roam Welty, War Bag Tyler, and the buffalo runners.

First Part: Westward, My Walden

"'Twas in the town of Jacksboro, in the spring of seventy three,
when a man by the name of Crego came stepping up to me,
saying 'How do you do, young fellow, and how would you like to go,
And spend the summer pleasantly on the range of the buffalo?'"
—*from The Buffalo Skinners*

First

The Better Part Of My Story began with the accidental death of an Italian boy, but first I've got to get you there.

Being well acquainted with myself and the persuasive power of my mother's tears, I crept out of my family home on Wabash Avenue in the middle of a sweet, firefly-lit April night with my shoes in my hands. I left a hastily scribbled note on the dining room table to assuage my parents' worst fears. I had only a wrinkled advertisement I'd clipped from the Chicago Sun to guide my way, but with it and a single piece of luggage (filled mostly with writing materials), I traded my meager savings (all culled from birthdays and my graduation) for a ticket and boarded a train.

That slip of newspaper led me to a dingy laborer's camp somewhere near Granada in the Colorado Territory, which was at that time the end of the Atchison-Topeka and Santa Fe line.

I didn't need to show the clipping around to find my way. The respectable passengers flaked away at each stop until only the most unwashed and unfortunate men remained. Together we stared silently at the blur of prairie through the yellow glass of the passenger car till the conductor called out 'last stop.' With few exceptions, we were all headed to the same place.

For the most part the men whose company I shared were foreigners; Slavs, Irishmen, Dutchmen, and an Italian not much younger than I. I single out this Italian because of the part he plays later and because I am sure he stole my watch as I slept on the bumpy wagon ride from the depot to the remote camp. When I reached the end of the long line of 'applicants' and brought up the theft to the pudgy overseer, the extent of the investigation went something like this:

"I ah . . . believe I've been robbed," said I.

"Of what?" said he.

"A watch. It was a gift from my father."

"Well, you don't need to know the time out here. Next!"

Having no particularly useful skills, nor even the wit to lie, I was delegated to the track laying team, hoisting ties and laying rails and doing my best to keep my feet out from under the blows of my partner's sledge. I never even knew that man's name. He had a lumpy red nose, fleshy lips, and did not respond to English. The ring of steel and iron was the only sound he ever made.

Thus far my great journey west had been a complete bust. I'd seen nothing of Indians, buffalos, or pistoleers. Just flat expanses of endless grass and gritty, coarse men who were underpaid and undereducated.

I tried to take a kind of romantic solace in the reassurance that each stroke of my partner's hammer meant the furtherance of eastern ambition. We track layers were not individuals (I wrote), 'we were fingers on the end of a long, Herculean arm that reached out all the way from Washington, sweeping away forests, diverting the course of rivers, and breaking down mountains like the desperate inebriate who knocks aside the place settings to feel the cool, curving reassurance of the bottle in his hands.' It amazes me now how brimming with bullshit I once was.

The level of it has gone down since then, a little.

Track laying was harder and more tedious than anything I had ever endured, and any sense of purpose was quickly driven out of me by sheer monotony. We were trapped in a mad race against the locomotive that puffed impatiently behind us, pulling along the cars of barely edible food, flatbeds stacked with heavy ties, and our own cramped, pest-infested sleeping quarters. Behind the engine toiled the blacksmiths and their assistants in the boxcar factory, churning out rivets and hammerheads by the bushel full.

I spoke little, and few spoke to me, except to barter in broken English for tobacco or food, or curse my clumsiness. The foreigners conversed in their tongues, but I was without a friend. The only books I saw were tally records and survey maps carried by the bosses.

I wiled away my strenuous hours plotting an epic confrontation with the Italian boy in which I would reclaim my watch and leave him cupping a bloody nose all the way back to the sleeping car.

It never happened.

One noontime there was a loud shout near the back of the line.

"Some damn fool fell under the cowcatcher," I heard a man say. We all ceased our work and crowded around the front of the engine to watch the idle men of the dynamite gang pry the body loose.

I had never seen a man in such a state, and turned away after only one gory glimpse. His body had been ground up by the heavy, ceaseless wheels, as if the black engine had tried to devour him. Only his face remained, like an uncorrupted saint's. It was the Italian boy.

A shadow fell across everything. It was black smoke from the stack of the terrible engine creeping across the face of the noonday sun.

I decided then and there to strike out from that abominable place, though to where and how, I had no idea.

Second

My Savior Came On A Shaggy Pinto one May afternoon. He was a Negro, one of the free lance buffalo hunters who sometimes came around to hawk meat and hides to the overseers.

I had seen hunters come selling tongues to the bosses in the caboose. The going price ranged from a dollar and fifteen cents to as much as five dollars. At these rates, ten buffaloes a day provided more money than I would see in a month of hard work. Never mind the hide, hump, and horns.

We had not seen a sign of outside life in days and my patience with the railroad had reached its limit. I had no knowledge whatsoever of buffalo hunting. I could not shoot, skin, nor prepare meat, and I preferred driving a wagon to riding on a horse's back whenever I could. Yet, my willingness to learn was directly proportionate to my desire to leave the rail gang. I could only hope that would be enough.

I had seen this particular Negro before. He was usually accompanied by a tall, thin white man with a thick blonde beard and curly hair. On this occasion, he rode alone. He was a rangy man about ten years my elder, with a quiet demeanor and ashy, walnut skin further darkened by the sun. His face was unshaven, but the beard there did not progress far beyond a wooly tracing around his lips and beneath his chin. He had a high forehead and big, deep eyes that glistened as though they would drop with tears at any moment

There were Negroes aplenty amid the railway workers, but what struck me about this one in particular was his mode of dress. He wore a faded, patched Union blue soldier's tunic, with the light marks of double arrow points on his sleeves the only remnants of the chevrons that had once ridden there. His hat was a gray felt affair with a floppy brim that hung down over his eyes at times, obfuscating them in a midnight shadow. A red bandanna hung around his neck. His breeches were threadbare, and his feet sheathed in high fringed leggings that ended just below his knees. A knife hung in a beaded sheath from an Army-issue belt around his waist, and there was a pistol in a flap holster butt first on his right hip. He had a short carbine in a scabbard on his saddle, and a kind of Indian bow adorned with bone tips and a gut string strapped to the cantle behind him. A quiver of feathered arrows bounced jauntily on the rump of his horse, tied securely with thongs.

I had never known a Negro by name, though I had of course dealt with them from time to time. It was not uncommon to see them downtown, running errands for their employers, or serving as bootblacks or sweepers. Father had always said they were as deserving as any other folk of liberty, though in his experience he had observed in them a tendency toward shiftlessness and irresponsibility, particularly where labor and money was concerned. This, he always said, stemmed from "the retardation of their moral and cultural evolution by their oppressive Southern taskmasters." Like father like son, I guess.

Had he been a white man, I doubt my approach would have been any easier. Yet he was the only course available to me, and I knew if I did not make myself known to him now, I might not get another chance for days.

I was in the midst of laying a tie when I let go of my share of the timber and walked away from the gang. As I approached the man on horse, the work boss blew his shrill whistle and shrieked after me. My fellows turned their heads to watch. I made no move to heed. I was through with them all.

The Negro was watching me too, from atop his horse a little way off. I marched over to him, my steps slowing as I got closer. I had no idea if he would be of any help to me, but I couldn't stay with the railroad another day.

I stopped in front of him, and his spotted horse snorted. I took a step back for its sake. The foreman was still shouting curses at me from behind. I felt like a beggar come to work for food. This was very near the fact, for I had none, nor the means to acquire any.

"You . . . buffalo hunting?" I ventured, and perhaps in slightly patronizing pidgin-talk, for I had the peculiar notion that he wouldn't understand me. My timid question was made further ridiculous by a little up and down pantomime I made with my hands, which I suppose I meant to represent a herd of buffalo running.

He smiled, a perfect row of white teeth that shone in his dark face like good china. He leaned forward, his hands crossed over the horn of his saddle.

"Yup," he replied, in a rich voice. "Lookin' for a skinner. 'Need a job?"

I looked over my shoulder at the furious work boss, red in the face and hexing my family for generations back. I turned back to the Negro hopefully. That made him burst into a laugh that could have shook an eagle out of its eyrie.

"Got a horse?" He asked, still grinning. "Got money? Ever been skinnin' before?"

I shook my head at every question, desperate for him to take me in, to take me anywhere.

He looked me over for an unbearable amount of time. I was sure I would feel the big hand of the boss jerking me by my collar back into my lamentable situation any minute.

"Well, we got a extry horse," he said. "But we goin' have to double up for awhile."

He reached one light palm down to me and motioned for me to swing on.

"C'mon. Mind that bow, now."

I gingerly took his sun-warm hand and clambered onto the pinto with no small amount of trouble, for there were no stirrups.

He struggled till I was finally situated behind him, and with a mortifying crack I heard and felt the handsome bow snap beneath my behind.

I muttered meekly in his ear,

"I'm so very sorry."

I fully expected to be flung down into the grass, but he just sighed heavily and told me to hold on. He urged the pinto into a trot and the curses of the boss were lost behind me, merging seamlessly with the unending din of the rail gang and the chugging of the engine before fading away like a bad memory. The only sound was the hooves of the pinto and the rhythm of its breath as it bore my new companion and I south over the tide of long grass.

There was a marked difference in seeing the prairie framed in the confines of a train window and in seeing it from the back of a loping animal. I could see forever. The wind was in my hair and ears, the grass was far below, hissing underneath . . . and the sky! My head was another cloud in the firmament, and my vision encompassed all directions. I was at once a part of the blue heavens and the rushing land and the methodically swelling flanks of the horse between my knees.

There was no conversation to be had the first lap of the ride, which lasted half the day. Though it seemed we rode with no purpose, I trusted the stranger to the extent that he plainly knew I had no money nor anything of real value. What little savings I had taken from home I had spent or had stolen from me weeks ago. When at last we stopped to rest and feed the pony, I told him my name and asked his.

"Roam Welty," he replied, wiping the lather from his pinto's back with the saddle blanket.

"I'm sorry about your bow. Was it an Indian bow?"

"It was mine," he said, though without malice. "Some tack in the bag, water in the canteen, if you thirsty. But go easy."

I was not used to the idea of a Negro who failed to punctuate every sentence with "sir." I knew of the existence of freeborn black men, but had taken it for granted that they spoke in the same deferential manner as all coloreds. I confess I was put off by Roam Welty's familiarity at first. Nevertheless, my livelihood was in his hands, and I saw no reason to strain my tenuous position.

He set to tending the horse, while I rummaged through the saddlebags. Between powder and lead, a dry bag of tobacco and one of cigarette makings, I came across something that felt like a small pelt. I took it from the bag to examine it.

I think I must have let out a yell, for Roam turned quickly on me, drawing out the long barreled pistol on his belt. I heard the hammer of the gun click back like the meeting of two malevolent fangs, and was sure he was going to kill me.

What I had taken from his bag was a curly, yellow-haired human scalp.

The grisly trophy lay at my feet. I had no weapon to grab, not even a piece of kindling. I silently laid a curse on the contemptible prairie for being bare of wood and thus my downfall.

Roam's dark eyes moved from me to the scalp and back again. I was sure some murderous cog behind his savage brow was clicking into place, just as the live chamber of his revolver had come into firing position. I felt myself thinking a thousand foolish thoughts. I wanted to blurt out that my father had been an abolitionist (Well, he was no John Brown, but he *had* supported that cause with monetary donations . . . occasionally), that I shared none of the usual misgivings of my race toward his. I wanted to point out that I was from the North, and clearly explain my family's utter lack of involvement in any trade that had utilized slaves or condoned the practice of his Southern oppressors.

Then the taut muscles in his face slowly relaxed, and his shoulders eased. He put the gun away.

"Where you from, mister?" he asked.

"Chicago," I answered warily.

He nodded, as if something had been made clear to him, and turned back to the pinto. The corners of his lips turned upward in a show of amusement.

"When you got the tack, jest put that back where you got it from."

I found a parcel of strange, dried meat and gingerly replaced the hideous fetish into the saddlebags and kicked shut the flap. Roam undid the tether on his horse, slipped it over his elbow, and sat down across from me.

I offered him his own food, without trying it. After the scalp, I was worried about its origins.

He took a piece and tore at it with his teeth, then motioned for the canteen.

I passed it to him, and he undid the top and sipped.

I watched him drink in silence. I was thirsty and hungry, but the thought of sharing a canteen with a Negro unsettled me.

I still had a piece of meat in my hand, and I turned it over, examining the spotted discolorations in what I had decided was dried beef of some kind.

"Washa," said Roam, watching me examine the food.

I looked at him blankly.

"Pemmican," he continued, around a mouthful of the stuff. "Dried buffalo with chokecherries. S'all they is."

I eased the tough meat in between my jaws and had a time tearing it free. Drool ran down my chin and in the act of supping it up, I inadvertently swallowed the piece whole. Lodged in my throat, it felt like a strip of tree bark. I tried to cough and found my passage blocked. I hissed soundlessly and reeled.

I felt my face turn hot and purple. My eyeballs bulged and my vision watered. Then Roam's wiry arms wrapped around me and squeezed. The piece of pemmican popped free and landed wetly in the grass. I fell on my side, wheezing.

My throat burned, and when Roam thrust his canteen in my face, I forgot my earlier hesitation and drank.

Roam pulled the canteen away then and sat back on his haunches, affixing the cap. He stared at me as I coughed a few more times.

"Damn," Roam said, and shook his head plaintively. "Mister, is they all like you in Chicago?"

I smiled, but wondered glumly if they were, or worse, that I was like all of them.

The rest of the ride Roam let the pinto canter easily so we could speak. He took it upon himself to teach me as much as he could before we met up with his employers. Doubtless the men he worked for would call his judgement into serious question when it became evident to them what a tinhorn fool I really was.

"If'n you ever get thirsty," he told me, "and y'got no water, tuck a pebble or a piece of rawhide in your jaw. If'n you find a buff wallow, you can drink outta that, but run it through a bandanna first, or some cloth. I call that 'buff tea.' It's bad, but it'll help you live.

'If'n you ever get lost, don't shout out, an' don't wander around tryin' to find nobody. You can't always tell what's the right way out here. Watch the sun, where it come up, where it go down. It don't never change. If'n you can, just shoot up in the air three times. Bang. Bang. Bang. Somebody around, they find you. Can you shoot?"

"I don't have a weapon."

"Well, but if you had one, could you shoot it?"

"I could shoot it, but I don't know that I'd hit much of anything."

"I'd give you Stillman's," he thought aloud, "but it's gone."

"I thought you wanted a skinner."

"Shore, but you never know. Indians and such. Snakes, skunks . . ."

"*Skunks?*"

Roam nodded, gravely serious.

"Shoot any you see. Can you ride?"

"I'm riding now . . . ."

Roam blinked.

"By yo'self."

"Oh. Well, I can drive a wagon," I said, earnestly.

"Well, that's awright, I guess," he said. "How about cookin'?"

We'd had a hired cook back home. A Polish woman. With luck, I could boil water.

"No," I said. "Not really."

Roam shook his head.

"Where was you raised, boy?"

My face flushed.

"Incidentally, who is Stillman?" I asked, to change the subject.

"Our last skinner," Roam said. "Stillman Cruthers."
"The blonde fellow?"
Roam looked at me sharply.
"I saw you ride in with him once."
He nodded.
"Yup."

He'd seemed like quite a striking character in his own right. I was anxious to meet him.

"Are we meeting up with him?"
"Hope not," said Roam. "He dead."
"*Dead?*"
"Indians kilt him 'bout three, four days ago. He takin' a shit, they rode up on him."

I wondered if my bowels would prove immovable out here.

"He was your skinner?" I asked.
"Most of the time he just drink," Roam smiled. "Sit around, read books to us . . . . Sometimes right while we was workin.' Woo! He get under the man's skin, thas' f'sho."
"What man?"
"Old War Bag. He top man. Don't never cross him. He a good enough ole man, but he get half-crazy sometime."
"How long have you been working for him?"
"I don't work for him," Roam said sharply.
I stiffened.
"I'm sorry. How long have you been with him?"
Roam's eyes softened again.
"S'awright. Goin' on two year now, but I knowed him long befo' that."

He glanced up at the sky then.
"Damn! Storm comin.'"

I looked, and thought I saw a darkening in the sky a good distance off. I couldn't believe it would get to us anytime soon.

"I don't think it'll hit us," I ventured.
"Think what you want, them storms move fast."

The preferred method of dealing with a storm on the prairie, I learned, was not to be caught under it. Roam kicked the pinto in the ribs and I had to hold on to his waist to keep from sliding off as we bolted off over the grass, the storm clouds still far out on the horizon.

Third

The Waning Spring Sky boiled over like gray soup when we arrived at our destination, an out of the way sod house dug right into a low, grassy hill. The crudely cut timber frame of the doorway, covered only by a stiff animal hide, was crowded with sloppily hewn, weathered shingles that bore various advertisements scrawled in a big, childish hand. They proclaimed "LIKKER," and "ARMAMINTS, TACK and CARDS," and on one shingle that hung completely vertical by one rusty nail, "KENO"—with the "O" on a separate square of hastily cut wood. Atop the pole and sod roof there was a large sign that read "CUTTER SHARPES." Recently, someone had added "LORD" in front of the big name. White smoke rose from a black iron stovepipe that poked out of the rounded roof.

There was a weathered log bench out front next to a rain barrel, and a dead tree trunk (the first tree I had seen since leaving the railroad) served as a hitching post. Five or six pack-laden, hobbled horses were tied to the knotty branches.

As we dismounted and tied Roam's pinto among the others, I spied a lone mule on a short rope tied to a cottonwood stump next to a woodpile behind the hill. The earth around it was completely bare. The mule had grazed down to the dirt, and now sat complacent in the middle of the mown patch, blinking at the plains of grass that were just out of its reach. I had never known a mule to eat grass before. Its ribs pressed out on its flanks, and I felt sorry for the poor creature. I moved closer, thinking to tie it up with the horses, as its careless master should have done.

A cracking voice spat at me from the doorway of the lodge, thick and blue.

"'Ere now. Wot is it you're doing?"

A silver haired, unshaven old man in a long underwear shirt and corduroy pants stood barefoot in the doorway. In his thin arms he held what looked like an old flintlock rifle, primed and ready to fire.

I raised my hands.

"I'm not armed," I said dumbly.

The Englishman squinted at me.

"Din't ask if you were. I asked you wot it is you're doing."

I looked at the mule and back at the man.

The Englishman raised the long rifle to his cheek.

"Gonna pinch my animal, were you?"

"This is your animal?" I asked, perplexed, staring anxiously at the big black eye of the old musket.

"It is. And why would you care to pilfer it, then?"

"I wasn't going to pilfer it. I was going to put it by the other animals. It's eaten all of its grass, and can't reach . . ."

"I know that, you impudent rascal," quipped the Englishman. "Wot of it?"

"It'll starve. It looks like it hasn't eaten in some time already," I said.

"It's my animal. Leave it be."

Roam walked up from the horses, his saddlebags over his shoulder, his rifle in one hand, and a long bit of wood broken off from the fallen tree in the other.

"How do, y'Lawdship?" he said to the old man.

The old man's eyes lighted on Roam, and he cocked his head.

"Is this bloody footpad with you, then?"

Roam nodded.

"Goin' be our new skinner."

"Well, keep him away from that beast. He has been sentenced." And with that, he lowered the rifle and ducked back inside.

I walked over to Roam.

"You mean to say *he* owns this place?"

"'Member I tol' you watch out for War Bag 'cause he half-crazy?"

I nodded.

"Don't cross Cutter, neither. He alla time crazy."

Roam stepped down into the doorway of the lodge and went inside. I was startled by his gall. I had never seen a colored so readily enter what seemed to be a white man's establishment. No placards pronounced Lord Cutter Sharpes' an institution which catered to his kind, and I was worried we would be in for a fight. Apparently the old Englishman knew Roam though, and had addressed him with a certain level of . . . it could not be deemed respect. Let's say . . . tolerance. But why?

The musty, smoke-filled confines of Lord Cutter Sharpes' den consisted of a large, open room crowded with a few old crates and milking stools for furniture, a stove in a corner, a plank and barrel bar to the side, and a hide partition with a shelf of trading wares beside it. The ceiling was low and covered with damp, spotted cheesecloth, supported by erratically positioned beams. The walls were reinforced with hickory stems. Lanterns hung here and there, and about eight furred and grumbling men sat about the low tables or leaned on the bar. A faded British flag hung on the wall. There were a few anti-royalist bullet holes in it.

The storm broke outside, and Cutter bustled about with an armful of rusty cans and pots, which he distributed amongst the men. I watched as they set the pots in seemingly prearranged spots on tables or in corners. Almost immediately there came a steady pattering as the rain water dripped through the covered ceiling and was caught by the deftly placed receptacles.

Roam led me to a table and introduced me to the other runners of our band.

My brother skinner to be was an immense Missourian with beefy arms and big bear fists encircled completely with a thick carpet of coarse black hair. He was a thumb over six feet five inches, and had great dark eyebrows, a tossed head of hair black as a crow's wing, and a curly beard and mustache which almost completely obscured his lips. A solid gut tested the buttons of his patched frontier blouse, which was nearly sheer at the elbows. His breeches were faded denim, and his big feet were wrapped in barbaric hide boots (I suppose because the commissioning of a pair of proper footwear for such gigantic feet would have broken the Missourian financially in two). One ear had been pummeled shut long ago, and he wore a fighter's nose several times broken. Though he had the look of a brawler, his black eyes were meek and calm.

He shook hands eagerly. Too eagerly, and much too firmly. He was introduced to me as Jack McDade, and he spoke in a careful, almost alien drawl.

"How d'ye do."

He was a tanner and hide man not long down from the Ozark Plateau, and incidentally, had swapped his old Kentucky rifle (the same one that had been pointed at me moments before) to Lord Cutter Sharpes for a jug of whiskey and a big bent Oriental knife the old gent had told him was from East India.

The outfit's star shooter was a lanky, self-important rogue with a mournful, stylish moustache, lustrous dark curls down to his shoulders, and striking blue eyes set in a tanned face. He was only a year or two older than me, but he spoke and carried himself like a man of seasons.

"George Pascal Fuchilarde *Latouche* is my name," he said, with such panache as can only be summoned by a man who has lived his entire life up to date with such an impressive moniker. "*Fuke*, is what they call me." His speech was practiced and baroque, evocative of white columned verandas creeping with ivy, and the swaying branches of old willows.

Fuke wore a wide pancake hat, a fringed, brightly colored blue and cream *capote*, and a pair of handsome, high heeled boots. Of Roam and Jack, he was surely the cleanest, and did not in the least look like what I pictured as a buffalo hunter.

"Are you from New Orleans?" I asked Fuke. My father's business partner, Jameson Carbuckle, hailed from that town, and I thought I placed the accent.

He ruffled at that.

"I'm no denizen of *that* filthy den of profligates and guttersnipes," Fuke spat. "My people come from *up* river, in East Baton Rouge (as though the distinction between East and West would mean something to a man who had never been there). New Orleans is the receptacle into which the refuse of *finer* cities—ah, such as Baton Rouge—is cast."

I apologized for my mistake.

I saw various figures at the crude tables. For the most part they were white men, with one tame Indian among them. They were all, by their slovenly costumes, frontiersmen. What stories must they have to tell, each and every one! I was able to discern some kind of card game going on between three men before I met the wild, bloodshot gaze of the Englishman, who was leaning over the bar watching me, I guess to make sure I didn't go out and make another try at his starving mule.

I quickly pretended to survey his wares. Roam caught me looking and told me not to worry about supplies, as they would outfit me with whatever I needed from the late Stillman Cruthers' gear.

I asked Roam about War Bag, but he shook his head and muttered;

"He be by later."

My new companions had secured a place near the stove and I took my seat. The smell of something tastier than pemmican wafted from a covered pot on the hot plate.

"Are all these men buffalo hunters?" I asked off-handedly while I pulled in my stool.

It was then that Fuke Latouche corrected my use of the term *hunter*.

"*Hunter*," said Fuke with evident disdain, "implies a perpetual *searcher*. A runner chases what he's after and *catches* it."

One of the other men in the lodge came from across the room to the stove and checked the contents of the covered pot. He produced a ladle and proclaimed whatever was inside "done."

The men all stood and gathered tin plates from the shelf on the wall. There were plenty to go around, and soon I had sat down with my new associates and dug into a plate of hot meat and potato stew. Cutter Sharpes was surprisingly hospitable considering his variable mental state. He gave us each a sourdough muffin peppered generously with raisins, which Roam called a huckydummy. He was the last to be served, though again, nobody seemed to take any offense at his presence.

After our repast, the absent leader of our outfit was still nowhere to be seen. Fuke induced me to try a portion of the crazy Englishman's "likker," and the concoction's spicy nature soon drove all imaginings of the mysterious 'War Bag' from my mind.
Roam leaned his stool against the hard packed earth wall and whittled away at the long piece of kindling he had taken from the fallen tree.
The day turned to early evening. The wind that had briefly sent oceanic ripples through the endless tides of grass floundered and died, leaving only a gentle but consistent drizzle of rain. The pans the Englishman had set up throughout the lodge beat a drowsy tune. The men took turns dumping them outside.
By that time my brain had been thoroughly drowned in the Englishman's elixir, and I vomited twice just outside the door. My stomach just couldn't abide the infernal concoction. Fuke sat snickering at my misfortune. Jack was fast asleep, curled on the floor like an oversized child clutching the wicked curved knife Cutter Sharpes had traded him as though it were a favored toy.
I drifted into a hazy slumber only to briefly awaken and once more empty my stomach of the Englishman's 'leprous distillment.' Cutter was calling out nonsensically in his sleep behind the bar. In my feverish state I felt sure that it was a woman's name he was calling, and that name was Dolores, my mother's own. Sickly, and in a superstitious frame of mind, I stumbled out of the lodge and fell into the wet grass.
The rain had stopped. I could still hear Cutter's muffled wail within, and I began to think of my mother back in Chicago. Was she ill? Had the shock of my leaving upset her health? I imagined her clearly, laying stricken on the floor of the dining room all day. No one would find her with my father at his business and myself away, my terse farewell note crumpled in her paling hand. The taste of bile and evil

rotgut was on my lips. In my drunken state, as I suppose all drunkards do, I wept for my mother and what tribulations I had wreaked upon her. I pined for home.

My heart was full of self-pity then, and in a sympathetic frame of mind, I found my way around back to the starving mule.

It was sitting there as it had been, drenched and wide eyed in the muddy bit of earth in which it had been left by its cruel master.

"There there," I said, or something in that vein.

I walked slowly up to it and stooped over to free the poor beast's fetters.

The damn thing nipped at me, bloodying my knuckles.

I cursed the animal up and down in a loud voice. I hopped about, squeezed my singing hand, and kicked a clod of mud at it. It stood up and began bucking, trying to catch me with its hind legs. I cursed it all the more, defaming its mother, its cousins, and every beastly ingrate of its foul strain until I was hoarse and the veins stood out in my neck. I decided to give the belligerent thing a good beating, and in a fit of besotted determination, turned to the woodpile and picked up a large piece of timber.

I raised it above my muddled head, ready to bring it down on the foul tempered creature's skull, when the bludgeon was plucked from my hands and thrown back on the pile in a clatter.

I was sure it was the Englishman, and was prepared to do battle with him simply for owning the pernicious mule. All I saw was a dark shape in the night which struck out and clipped me hard across my jaw, chipping a molar.

I slumped to the wet earth. The same shadow towered over me. In the dim moonlight, I saw that what I had mistaken for a pair of massive shoulders was in fact the carcass of an antelope, propped over the shoulders of my assailant. The dark eye of the dead animal glistened, and its pink tongue protruded from between its black lips.

I rose up on my elbows and met the cold muzzle of a rifle. It pressed against my breastbone, hard and insistent of its authority. I did not argue, but lay down flat on the grass and closed my eyes, not wanting to see the flash that would kick me into Hell.

I love my mother, was the last thought I had before I drifted into a pleasant, crapulent sleep. I was told that I began to snore, and with Homeric enthusiasm.

*   *   *

A tear from the sky splashed on my closed eyelid, followed quickly by another. I had been left where I'd passed out, lying on my back in the wet grass. My clothes were soaked, and I was shivering. My mouth was dry, and a heavy ache permeated my skull. I slowly sat up. The starving mule looked over at me with lazy disinterest.

All about the wide open prairie the heavens were spilling. The grass waved as the dark clouds wrestled and flung thunder over the land. It was early morning and the sky was blue-gray, mournful and bigger than I had ever seen it.

I heard voices in the low earth lodge. I stood up, hugging my elbows for warmth, and scurried inside as the weeping rain became a stinging spatter of full-blown lamentation.

A game of euchre was underway, with Fuke and Jack on one side and a gritty looking red head and a smallish Slav on the other. Jack, who was so huge he had to stoop in the sod house even while sitting, was drinking more devilkill whiskey and chewing tobacco all at once.

"God," I said, at the sight of it. "How can you do that, Jack?"

Jack looked up innocently, unsure as to what exactly I was referring.

Fuke shrugged.

"Well, he has to," he said. "Y'see, *Juniper,* (and I ruffled at that bit of condescension) ol' Fat Jack has got a very delicate condition. Circleptilaphobia. He's *deathly* afraid of hoop snakes."

"Hoop snakes?" I asked.

"Sure," said Fuke, and the others at the table looked from him to me, their faces mild as if it were common knowledge. "A hoop snake is a kind of Missouri rattler. Lives in the mountains. *Hates* tobacco. If you spit chaw in their eyes, they leave you alone. But they're *deathly* poisonous. Long as you stay out of the mountains, you're usually safe. But every summer, right around now, they make a *big* move to the prairies on account of all the jackalopes."

He shuffled his cards back and forth in his hand, still looking up at me. "It's *quite* a sight. They bite their tails and roll down out of the mountains in *packs* of about twenty or more. You can usually hear them coming though, like a bunch of Mexicans shakin' *maracas.*"

Fuke had me rapt until the very end, when I shook my head. All the men at the table burst out laughing, save for Fat Jack. He jumped a little, and from then on looked about nervously, sipping his whiskey more slowly, as though he had missed something important.

The bit of fun at my expense over, the men returned to their game.

I looked around for Roam and found him in a corner by himself. He had taken his broken bow and salvaged from it the bone tips and string. Now he was busy fashioning the bit of wood from the pile into a serviceable replacement.

I ambled over, as he was nearer to the stove and I wanted to get warm.

"Any word from War Bag yet?" I asked.

Roam nodded to the bar, speaking around a mouthful of something.

"There he is, right there."

I saw the back of a man in a torn brown hat and a patched, tobacco-colored corduroy coat with a wool collar hunched over the bar, talking with Cutter Sharpes. A scraggly plait of bone white hair dangled between his shoulders like a rope of carved, yellowed ivory, the tip dipped in tar. He wore a canvas cartridge belt cinched tight around his waist, from which hung a long knife and a hide tobacco pouch.

Sprawled on the plank like a pagan offering was the antelope carcass. Cutter was cleaning it as they spoke. It stared blankly as a toy while the Englishman ravaged it.

"He came back last night," Roam said. "Almost kilt you for a mule thief. Don't you 'member?"

I remembered being struck, and felt the chip in my back tooth with my tongue. A part of me was gone forever.

I looked back at the man at the bar, and slowly sat down across from Roam.

"What did he say? Where was he?"

Roam shrugged.

"Ain't said nothing yet."

I sat and watched Roam bend and shape his bow. Every so often he picked up his knife to make a correction, gluing on shaven layers of wood to strengthen it. I tried to ask him questions, but he had a mouthful of braided sinew.

Fuke and Fat Jack remained at the table amid the occasional burst of raucous laughter or bitter curses.

The man in the dull brown hat who was called War Bag did little but drink. He produced a wooden pipe and filled it. Thereafter, he alternated between speaking lowly to Cutter Sharpes, smoking, and tipping whiskey. He did not turn around, so I did not see his face.

"Where did you learn to do this?" I asked Roam, as he wrapped the new bow in strips of hide and canvas.

"Down in the Nations, livin' with the Muskoke," he said. "When I was little, we used to fish with these off the banks of the Deep Fork. Tie a string to the end of the arrows, drag 'em out floppin' on the bank." He smiled, seeing the fish kicking at his bare feet long ago.

I thought to ask him if he had been slave-born, but remembered his flash of annoyance when I'd suggested that he worked for War Bag, and thought better of it. The cavalry tunic was draped over his chair, and I turned my curiosity on that.

"Were you in the Army?"

The answer came from behind me in a singular, commanding voice.

"Roam was with the 10th Cavalry. He was one of the best trackers Carpenter had."

I turned, and War Bag looked down at me and extended his hand.

"Ephron Tyler," he said. His voice was rich and full of bass, enamoring. It was like thunder tumbling down in a deep river valley, but it could also be the low threat of an old bull that has been whipped for the last time.

Ephron 'War Bag' Tyler got his name because he was like what the plainsmen called a *war bag*; a stiff, hide parfleche, worn and tough. His coarse, wild, yellow-white sideburns framed a jagged, minimal face from which the skin hung like leather drapery. His neck and cheeks were dusted with perpetual gray-white stubble, like frost that thickened, tapered, and split into a queer fork on his chin. He had a slanting moustache over a wide, mean mouth, and a slightly hooked nose. His eyes were calming, dark, but radiating assurance. His eyebrows were jagged white lightning, the left one split by a faded pink scar that trickled down from under his hat like a meandering gully and petered out just at the bridge of his nose, marring a face that was otherwise grandfatherly, lending it a cruel air.

His expression was stoic and slightly disapproving of everything he saw. The pipe clamped between his jaws added to the air of authority.

I took his hand. It was the branch of a gnarled tree, hard and unyielding. I gave him my name.

"You ain't no Jayhawker," it was more of a statement than a question. I wondered at the sound of his voice. How raw I imagined his throat to be from the impossibly low sound that issued from it. Every intonation was rich with flavor, every pronunciation familiar, as though the old voice had spoken the words already a thousand times before.

"Where you from? Illinois?" He did not say 'Illi-noise,' as most non-natives did.

"Chicago," I said.

"I'm a Hoosier, myself," he admitted around the pipe, pausing to leak smoke. I spied his stained teeth gripping the nicked stem. "Boonville. What're you doin' out here?"

"I was working for the railroad," I said.

"Damn railroads," he growled. "Gonna be the death of this country. Drag along pimps and whores and worse goddamn things wherever they go."

He eyed me, and I thought he would ask if I had ever worked any of the professions he'd mentioned. Then he looked to Roam.

"I picked him up in a bad way," Roam said. "Goin' be our new skinner."

War Bag turned back to me, appraising me. He took the pipe from his mouth.

"We don't need no new skinner, Trooper. Take him back to his railroad."

A lump caught in my throat at the thought of being deposited back on the rail gangs, when there was so much promise here of what I had come seeking.

"We goin' need a body to replace Stillman," Roam said. "He do well enough, I guess."

War Bag never stopped looking down at me.

"We don't need nobody. Stillman hardly did any goddamn work as it was. Anyhow, we got better things to do now than run buffalo."

Roam sighed heavily, as though this were a rehash of an old and unwanted argument.

"Ain't goin' be nothin' but runnin' buff for me, Eph. Them men signed on with you to make cash money, not hunt no pretend Indians."

War Bag turned to Roam and put his knuckles down on the table with a bang, making me jump.

"Pretend, is it? Well I'm tellin' you, I seen Stillman rubbed out, and I seen the one that done it! Ridin' a big gray mare, he was. Sighted their trail about eight miles northeast of here . . . ."

With that, Roam reached into his saddlebags and threw the blonde scalp I'd previously discovered onto the table.

"That there is Stillman's scalp," Roam said.

War Bag quieted down. He put the pipe back in his mouth and picked up the scalp, turning the thing over in his hands. Several of the other men inched closer to see it.

"The Dog Soldier I took it off weren't no powder burned Indian," said Roam. "He was a young buck on a spotted gray stallion, black paint on his face."

War Bag murmured something low, and his eyebrows kissed at the sight of the scalp.

"'Rest of the war party split off and forded the Arkansas. They probably followin' the Big Sandy up a ways till they steer for Nebraska," Roam said. "From there they most likely get lost in with the Northern Cheyenne."

"They might've traded horses," War Bag muttered.

"Not likely," Roam said. "They had they selves a whole string of shod ponies they got raidin' down along the Carrizo. That gray was unshod, and it's left foreleg was fin'a go lame in another day from a piece of shale it picked up by the river. Been with the band awhile, weren't worth tradin.'"

War Bag looked at the scalp for a long time, then let it drop on the table between us.

"Maybe," he acceded.

"I thought the Indians were all tame in this part of the country," I said.

War Bag stared at the scalp on the table.

"Lotta country and a lotta Indians. Can't all be tame." I wondered if he meant the land or the Indians. "Indians know most every nook that'll hide man or horse. Cheyenne Dog Soldiers are the worst kinda Indian."

He fingered the pink scar on his eyebrow unconsciously as he spoke.

"Dog Soldiers dyin' out," Roam said. "Most of 'em reservation Indians now."

"Tell Stillman," War Bag said, before he turned his eyes on me. "What about you? Ever been a skinner before?"

"No sir," I replied.

"Helluva time t'have to learn," he said.

Fourth

In The Sod House, strips of antelope meat were roasting and the euchre game ended as most card games do, in hard feelings. Fat Jack and Fuke wound up deep in the hole, and the Slav and the red head congratulated each other loudly.

I decided from watching War Bag and the way he and Roam interacted that Roam's acceptance in the company of the other white men had something to do with his relationship to the old man. In all things the others deferred to War Bag. Jack and Fuke naturally yielded to him as their employer, but even the men of other outfits seemed to unconsciously respect him. He was given ample space at the bar, and even the youngest and most brazen of the runners were not wont to cross him.

War Bag treated the others in a paternal fashion. He quietly listened, sometimes smiled, nodded as others spoke. But he only had words with Cutter and Roam. These two were on a level closest to his own. With Cutter, for the most part, it was his age. But with Roam there was something more elusive. The regard with which the old man held the Negro seemed to assure his acceptance among the others. What had War Bag done to command this position among the others, I wondered? And what had Roam done to so win his respect?

"And why in the *hell* are we taking this greenhorn *juniper* along?" Fuke said. He was in a bad mood over losing to the Slav and the red head.

"He ain't got no place to go," said Roam, "and anyhow, he got a quick head and good timber."

"A quick head won't stop a bullet, or a *damn* Indian arrow."

"They ain't goin' be no wild Indians where we goin.' Just big bull buff and skinny calves on the teat," Roam said.

Fuke turned to me.

"A bull buff'll run you down as quick as an Indian, *Juniper*," he goaded me, and stared in my eyes, a little drunk. "How 'bout that?"

Everyone in the room turned to watch, and I felt their eyes. I could feel too Fuke's hot, candied breath beating on me.

"I guess I can outrun a buffalo," I said, as calm as I could.

Everyone laughed. Even War Bag, from his place at the bar.

Fuke cracked a smile and stepped back.

"You damn fool. I've seen *horses* can't outrun buffalo."

"I guess I'm not a horse," I said, still even, and smiling now. "I'm a man. A better one than you."

The laughter died down, and everyone looked grave, glancing between the two of us. I thought perhaps I had gone too far, but Fuke folded his arms and made a big show of humoring me.

"And how'd you come to that *astoundingly* stupid assumption, boy?"

"I'm telling you I can outrun a buffalo. You said yourself you can't."

"I didn't say that . . . ," Fuke began.

"*Could* you?" I interjected.

"*Nobody* could," he said. "Not even an Indian." He jabbed a finger at me. "And *not* you."

Fat Jack spoke up from the table, where he had been quietly listening.

"How would we know?"

Fuke looked down at him.

"What're you talkin' about, Fats?"

"Well, how would we know if'n he could or couldn't? Thar ain't no buffalo to race 'round here," Fat Jack said.

Fuke shook his head at Fat Jack.

"Don't be *stupid*, Fats *nobody* can outrun a buffalo on foot."

Fat Jack looked at me and shrugged.

"Well, I reckin thar ain't no way to tell."

Some of the men listening at one of the nearby tables got into a discussion. Then the red headed euchre player spoke up.

"Hey, I might know a way."

The red head stood up and pulled a reluctant man from the seat next to him. It was the tame Indian I'd seen before. He was about Roam's age, grim and thoughtful, with the thin, sparse makings of a moustache over his lips. He was dressed in an old Rebel infantry uniform, and sported a beaver hat. He looked extremely uncomfortable to be the center of attention.

"This here man," said the red head, "is Caddo Charlie. He was a messenger for the stars and bars and he's the fastest feller I ever seen. He can sup fifty miles in a day. If anybody can outrun a buffalo, he can."

I looked at Caddo Charlie. He was lean and tall, with the long legs of a good runner. I had been a top half-miler myself in school, and had been goading Fuke with the notion of racing him, for I was sure I could beat him. But hearing

Caddo Charlie's feats so vaunted, combined with what I knew (or thought I knew) of Indian tenacity, I was worried that my little boast had gotten me in over my head.

The gathered hunters erupted into a frenzy of wagering. Cutter Sharpes was elected to the dual office of bookkeeper and treasurer. A pile of cash money quickly appeared on the plank bar, and even War Bag had to move his drink aside.

"Ain't hardly no place to run," said Roam.

"What're you *talking* about," quipped Fuke. "We got miles of grass all around—not even a *prairie* dog hole to fall in."

"*Wet* grass," said Roam. "Ain't hardly the right kinda field to have a runnin' race on."

Fuke stooped over and dug deep in his boot, retrieving a small cache of dollar coins in a red pouch.

"You son of a bitch, you said you was busted," hissed the red head at the sight of the money.

Fuke shrugged at the man and placed ten dollars on the bar, letting each one resound on the plank as he dropped it, so we all could count them. He tipped his hat to Cutter Sharpes.

"With your permission, *your Lordship*?"

"Men have raced horses, dogs, even mules in front of this place," said Cutter. "But a foot race? Is this a schoolyard, then?"

I watched as the red head spoke a few quick Indian words to Caddo Charlie. The Indian nodded, took off his hat, coat, shirt, and boots, and stepped outside, not even sparing me a glance.

I followed him, and behind me went the entirety of the house, with Roam and War Bag bringing up the rear.

The sun was shining down on the grass, which was slick and treacherous with frequent patches of mud. I took a hint from the Indian and took off my shoes and socks. My feet sunk into the damp earth. The cool feel of the mud and the grass blades through my toes was pleasant, but the damp ground sucked at my feet as I walked. That would work against me.

There was a rawhide cord with a small bag dangling from Caddo Charlie's neck, and he reached up and broke it. He hunkered down, drew out a pinch of something, and took it into his mouth. He chewed it for awhile, then spat into his open palm and rubbed the brownish juice on the bottoms of his feet.

The gathered men murmured in appreciation and awe.

I stretched my limbs as I had done before every track and field meet I could remember. That got quite a laugh out of the men. Old War Bag's tremolo guffawed above the rest.

"Hey, he's got his own runnin' medicine!"

"This is a *foot* race, not a ballet!" Fuke guffawed.

I ignored the slapping knees and fool whoops that sounded all around me.

Roam had his new bow in hand. He fitted an arrow, arched it up, and let fly. The arrow flew a while through the air, then plummeted into the grass where it stuck in the earth, its pintail feather shaft quivering.

Some of the men whistled at the distance.

"Hell, I'd count myself fortunate to make it out there, let alone beat another fella runnin,'" said one man. The others laughed again.

Caddo Charlie closed up his medicine pouch, raised it to the four directions, and fastened it around his neck again. His skin shone liked red clay in the sun, the muscles underneath rolling easily. The tendons in his bare legs stretched, ready to propel him through the grass. His anointed feet stirred, anxious to go.

Roam looked at me.

"There and back. When you ready," he said.

I rolled my neck on my shoulders and touched my toes. When I came up, I said,

"Ready."

Roam looked at Caddo Charlie, and the Indian nodded and braced himself to spring from a wide stance.

I got down into the classical starting position—bent all the way over with my right leg stretched back. All my weight forward on my left leg, and my fingertips stretched out and touching the muddy earth before me. I was like a catapult ready to let loose. The hunters hooted and made kissing noises.

"He gettin' ready to run, or get poked?" roared the red head.

"I wouldn't do that 'round here, *Juniper*," said Fuke. "Some of these boys ain't seen a lady in *many* a year . . ."

"*GO!*" Roam yelled suddenly.

Caddo Charlie leapt forward and was off like a cannon ball.

My left foot slipped in the wet grass and I stumbled after.

The hunters became one voice, cheering and hollering. They were all drowned out by the beating of my heart as I raced to catch up with the Indian, who was looking back over his shoulder at me as he ran.

To my surprise, he actually slowed his pace a bit, looking confused. I gained a few feet till I was nearly behind him, and he turned back to the business at hand and gave me the race of my life.

The grass and mud below made the run taxing, and I had to exert myself double just to keep the pace. I knew of course that the run to the arrow would not be as important as the run back, but at the rate the Indian was going now, by the time I reached the turning point the red head would be pouring a celebratory glass of whiskey down Caddo Charlie's throat back in the lodge. It would surely be back to the railway gang for me then. Imagining that I was running away from that fate, I think I found new speed.

I pumped my arms and pistoned my legs, my eyes on the arrow shaft as it grew larger. To my right Caddo Charlie dodged something that squeaked and scurried away in the grass, and he slid for a few feet on his heels, madly swinging his arms to regain his balance.

I gained on him just as he did so, and we reached Roam's arrow at the same time, digging our heels in to turn about.

Right beyond the arrow, where no one had seen it, a shallow depression that had once served as a wallow for some overheated herd animal appeared before us. This time we both slipped, and for a second we were fully parallel to the ground, frozen for one bizarre moment in mid-air. Then we splashed into the muddy puddle and scrambled through the dewy grass, shoving unintentionally at each other in a desperate attempt to regain our footing.

I had to wipe mud from my eyes as I clawed the ground and pulled myself up and over. Then I was stumbling clumsily alongside Caddo Charlie.

This was when the race truly began.

I could no longer separate the roaring wind in my ears from the shouting of the hunters. The grass whipped my bare ankles as we tore through the field toward the lodge, the big 'LORD CUTTER SHARPES' sign looming before us. My breath came out in labored huffs. My legs were hungry animals, eating distance.

Caddo Charlie was right alongside, his long hair whipping behind him. His face was like that of a diving hawk in the wind. His white teeth flashed in his mouth and a shrill sharp cry began to burst from his lungs with such force that I was sure he would collapse from the exertion.

Then we were abreast of one another. Our heaving shoulders brushed and withdrew like the testing foils of opposing duelists. He was the best runner I had ever seen, then or since. As graceful and swift as an antelope in flight before a fire.

We crashed into the mob of waiting spectators. Caddo Charlie bowled into the Slav and Fat Jack, knocking them flat. I flipped right over the bench and upset the brimming rain barrel, drenching Fuke from the waist down in last night's rainwater.

There was a moment as the Slav, Fat Jack, and Caddo Charlie dazedly picked themselves up. Fuke cursed me and kicked at the empty barrel that had soaked his breeches. Many of the men laughed and clapped me on the back as I was hoisted to my feet. It seemed to me that the race had been so close no one would be able to declare a winner.

Then Caddo Charlie broke into a torrent of rapid Indian-talk. When he finished, he slapped my shoulder and shook my hand. Then he strode off by himself with his arms spread wide and howled at the empty prairie, seemingly in anger.

The red head took me by the arm.

"Charlie said he bought his runnin' medicine from an Assiniboine, and it ain't never failed him. He says you must have lived with the Assiniboine because you knew that the only way the medicine would work was if you first run in your opponent's tracks. You let him get ahead of you, so you broke it." The red head laughed. "Didn't take you for part Injun, Stretch!"

I smiled thinly, and watched Caddo Charlie bawl at the prairie. He tore the medicine pouch from around his neck, threw it down in the mud, and proceeded to stamp on it.

So it was that by my opponent's admission, I had won the race. What better judge could there have been?

I was led into the lodge with much fanfare, and a jug of Cutter's whiskey was opened, passed around, and drained. Cutter dished out the antelope and more potatoes. It was the best meal I'd had in a month. By evening I found myself with a pleasant buzzing in my head, and a warm, solid lump in my stomach. It was a good change from the nausea and delirium of the night before.

The buckskinners warmed to my presence. I was presented with a few tokens; a buffalo hide coat for next winter, a tin pan and iron fork, a wooden canteen, a mirror for shaving (but no razor), a tobacco pouch (but no tobacco), and a big brown wide-brimmed hat that brushed at the tops of my ears. Lastly, Caddo Charlie, with much ceremony, presented me with a beaded Indian parfleche, replacing the bag I had left behind when I fled the railroad.

The red head produced a battered fiddle and began to play a filthy ditty called *'The Whorehouse Bells Are Ringing.'* Fat Jack accompanied on a Jew's harp, and the Slav brought out an old mandolin. Then, to my surprise, War Bag sang in a low but raucous voice that was glorious and intoning, despite the irreverence of the lyrics:

> *"The whorehouse bells were ringing,*
> *while this pair's upstairs in bed,*

trying to get their guns off first,
Into each other's head . . . ."

The others joined in, all save for the Englishman, who wandered outside and stood in sight of the doorway, watching the sun sink.

Roam passed me a portion of the money he had won (he had been the only one who wagered on me). He told me in the morning he would direct me on how to best spend it in further outfitting myself. It amounted to about forty dollars.

Only Fuke remained aloof towards me, and stood with his thin bare legs poking out from under his shirt near the back of the lodge. His drying britches hung over the stove, and he would not accept any drinks, nor allow himself to be drawn into the revelry. He glowered at me when I chose to look at him. Mostly I did not.

Later that night as the hunters spread their bedrolls on the floor, and Cutter Sharpes retired to the curtained bedroom off the main room for another night of tormented sleep, Roam gave me Stillman Cruther's gear.

Among the dead man's things were a big bone handled 'Arkansas Toothpick,' a red wool scarf, a kit of awls, pliers, and cutlery Roam said were used for skinning, a colorful Whitney blanket, and some saddlebags. In the saddlebags I found a few books belted together in a bundle, including a dog eared edition of Shakespeare's Complete Works, two rags (one the January edition of *Harper's Weekly*, and the other an old yellow copy of *Seth Jones*), and a copy of Melville's *The Confidence Man,* which I was anxious to read.

When Roam saw the enthusiasm I displayed at discovering the books, he shook his head.

"Lawd, not another one," he said.

I decided to ask Roam once again about my predecessor.

"'Still was the smartest fella I ever knowed," Roam said. "I think he might've been a schoolteacher once. He could talk Shakespeare without lookin' at the words, and he could cut a man down with his tongue like anything. But that didn't save him. Out here a man got to have more than jest book learnin.' 'Still read them there books and tried to be like the folks in 'em. That's why he just a dollar size hunk of hair now."

I shivered, though it was not cold.

"How well did you know him?" I asked. "Did you travel together long?"

"Not long," Roam said. "I think War Bag knew him a little longer. Go to sleep, now," Roam said, and rolled over.

But I did not. Not for a long time.

Fifth

We Set Out from Cutter Sharpes' early the next morning. City folk tend to overlook a thing like the dawn, but they are as diverse as snowflakes. In all my life there has never been one quite alike, and there never was another like the one that greeted me as I stepped out of that sod hovel my first day as a buffalo runner.

After buying what seemed to me a great amount of hard tack, a good saddle blanket, and a one sided file for the upkeep of Stillman Cruthers' (for I was not yet used to calling anything mine that had recently belonged to a dead man) skinning tools, I was left with sixteen dollars.

That last sixteen dollars went to the purchase of my first firearm, a twenty year old ten shot .38 Volcanic pistol. Roam advised me that the gun was obsolete, but still a better buy and a little easier to use than the hefty Walker Colt, which was the only other pistol Cutter Sharpes had to offer I could afford. He advanced me a portion of my wages and bought out the Englishman's supply of ammunition, one dusty, half empty, hundred round box.

The Volcanic was a queer, boxy thing somewhere between a pistol and a lever-action repeater. The brassy finish was tarnished and scratched. It had a six-inch barrel and seemed like a pocket Hotchkiss gun to me. Yet there was something enticing about holding it and knowing it was mine. Ugly though it was, it had a timeless charm. It did not quite belong anywhere, and would have been readily at home thrust through the sash of a buccaneer, or the belt of Natty Bumppo.

I had no holster, so Roam advised me to put it someplace dry where I could get at it. I procured a long rawhide cord from somewhere and hung the pistol from my neck, easing the barrel behind the band of my trousers.

My final gift from Stillman Cruthers was his horse, a fidgety chocolate shavetail named Othello. Roam's own pinto was called Crawfish, which he told me meant the horse liked to buck backwards and high-tail it when frightened. War Bag rode a giant broom tail black named Solomon that dug at the earth with his forelegs, anxious to go. Fuke had a fine Appaloosa he called Napoleon (after the song, not the Emperor). Fat Jack had a flea-bit gray ('Foggy') that was barely more than an old plough horse, and shied away whenever the Missourian saddled him.

Roam laughed at my insistence on knowing the names of everyone's horses.

"They ain't dogs," he said. "Ain't goin' come when you call 'em."

War Bag announced that we would strike out for Fort Lyon to hire a pair of drivers and wagons and requisition some lead. From there we would hunt in east Colorado and west Kansas, and winter in Dodge City.

"Kansas?" said Fuke. "Kansas is played out. The *panhandle's* the place to go, and the future's in *hides*, not humps."

"That might be," said Roam, "but they's a treaty with the Comanche and the Cheyenne sayin' we got to stay north of the Arkansas."

"I never heard of any *Indian* keeping a treaty," Fuke muttered.

"No white man neither," muttered Roam.

War Bag said "Hua!" to his horse and went off without another word.

I turned in my saddle and saw the thin gray mule behind the lodge. It was lying on its side now, and I could see its ribs swelling and deflating.

"Why is he starving that poor animal?" I asked, shaking my head.

"That poor animal kicked his wife in the head about a week ago," rumbled War Bag. "Old Dolores Sharpes was a strong woman. She lasted nearly three days."

Cutter Sharpes himself came out of the lodge and sat down on the bench. He was sunken and old as I looked at him now, like the mule.

Soon the trading post, Caddo Charlie, the red head, the Slav, and the mule all became one dot in the breezy grass.

We rode for hours in relative quiet. Fuke dozed in his saddle, leaning far toward the animal's neck, but never fell off. Once his horse stopped to graze, and being in the front of the line, War Bag and Roam rode on ignorant. I stopped Othello, thinking to wake Fuke up. As I watched, Fat Jack silently took the reins of Fuke's horse and guided the animal along behind his own.

The big Missourian caught my glance and shrugged, almost bashfully. I decided he was pleasant enough. I still had no desire to share his close company, filthy as he was.

The land grew more barren, and when we found the Rule River and followed its winding course, the going was rocky. The sound of the horses' hooves scraping the stone had a nostalgic effect on me, and I recalled the wheels of carriages on the cobbles of home.

A little after noon we watered and fed the horses, and Roam showed me how to wipe down my mount.

Fat Jack took off his boots and stepped off the bank into the water, looking up and down its length.

"'Reckin the water's cool enough," he said.

"Shoot, this is no *river*," Fuke said disdainfully, as he led Napoleon to the bank. "Mississippi, now *there's* a river."

Fuke picked up a flat stone and sent it skipping across the water.

War Bag filled his pipe and smoked, thoughtfully scanning the terrain. The smell of his pipe tobacco reminded me of my own father's habit of smoking in the yard, or in his study while he read. My father had an ivory pipe with a whimsical face carved in the bowl, and when he smoked I thought of his dark stockings poking out from beneath the cuffs of his trousers whenever he sat.

Roam slipped a feed bag over Crawfish's ears.

I hunkered down and raked my fingers through the smooth stones beside the river. Overturning one, I found a fossilized bit of coral. The stone was dotted with sucker-like marks, as though an octopus had gripped it and imprinted its tentacles. I looked across the land and thought of it filled with sea water. I thought of sharks or other, unimagined monsters hunting the floor of their benthic domain for some fearful school of fish, just as we would hunt the buffalo. Maybe the land rolled now like the waves that had once traversed it.

I put the fossil in my pocket.

Fuke reached over to his saddle and drew out his Winchester without warning. Before any of us could act, he had cocked it, drawn it up to his cheek with a little flip of the wrist, taken quick aim, and let off a shot that cracked out across the river and clipped a jackrabbit dead in mid leap.

I was amazed by this, but War Bag and Roam seemed unimpressed.

Fuke looked at Fat Jack.

"That'll make a *fair* stew, eh Fats?" Fuke said to the barefoot giant.

War Bag stepped up beside Fuke and whistled low. He clapped Fuke's shoulder.

"You bet. Now you just swim on over t'the other bank and get it."

Fuke lowered his rifle, and Roam burst out laughing.

Fat Jack looked at the two men on the bank, and back at Roam. He looked a little confused again, and started rolling up his pant legs.

"Don't fret, Fuke," he said. "Ah'll fetch it."

Roam shook his head, still smiling.

"Jest leave it, Fatty."

Fat Jack stopped, a little uncertain. He lumbered out of the water and put on his boots.

Fuke silently put his rifle back in its scabbard, glaring once at Roam.

"It was still a good shot," I said. It had to have been at least seventy yards.

Fuke spared me an evil eye.

We mounted again and continued on, and I rode beside Roam and asked him if we should be worried about another storm.

"Naw, that was the first rain I seen in a while."

"It's usually pretty dry," said War Bag, from my left. "Goddamn grasshoppers near ate up the whole world last summer. Them grangers had a helluva time." He chuckled lowly. "Serves 'em, I guess. Tearin' up the country with their ploughs and bustin' up the land. A man ain't meant to sit on a plot of dirt and watch it sprout. It's a piss poor way to live. Land throws everything at you. Heat, wind, varmints, hail, goddamn insects . . . . Hell, even a dog don't need to get beat twice."

Fat Jack heard a snatch of War Bag's talk and came up behind.

"My paw raised taters an' a patch o' corn back home," he sighed, his eyes smiling. "Come spring time, we'd brew up mash whiskey down in the holler. T'ain't nothin' like it."

War Bag said, "You're right about that."

"Why'd *you* come out here, *Juniper*?" Fuke said, for he had ridden up as well. "Squander your *inheritance*?"

"I want to be a writer," I said, dreading but fully expecting Fuke's drollery.

Instead it was Roam who snorted.

"Sheeyit."

"You mean like a newspaper man?" War Bag asked.

"No," I said. "I mean like a book writer."

"Books! Like make-believe? Writin' ain't worth a damn outside o' the newspaper."

I was stunned by this reaction from the leader of our group. What sort of ignorant people had I fallen in with?

"There's more to life than what's in the newspaper!" I exclaimed. "What about history and . . . and . . . dreams?" I said, feeling foolish erupting into artistic indignance on a Negro and an old frontiersman. "What about people's thoughts? Why, books can change a man's whole philosophy—his whole way of thinking!"

"A man's thoughts ought to be his own, 'less his opinion's asked for," said War Bag, not even ruffled at my change in pitch. "And any man that's gotta read another man's words to make up his own mind ain't worth a jughead mule."

Roam just shrugged, scratching his chin.

"What about the Bible?" I said, desperately trying to appeal to their common Christianity.

"No gospel sharp ever convinced me it was worth readin,'" War Bag said.

"But your horse's name is Solomon!" I said, exasperated.

War Bag looked at me sideways.

"For my uncle," he said.

I looked at Roam.

"I could sign my name," he admitted. "That's alls I needed t'get by so far."

"Ignorant, godless *heathens*," said Fuke, of all people to argue my point. "I'll say one thing, *Juniper*. Living in civilization gives you an appreciation for the *higher* things in life, but it tends to put you at odds with the ah . . . *provincials.*"

"Fasten your lip, Louisiana," War Bag growled. "And you, Stretch," he said, addressing me. "You keep your ears and eyes open and your goddamned books shut, and maybe you'll learn somethin' important."

It turned out Fuke had some degree of education, though his vehement dismissal of certain authors made me think he had limited experience or taste. He absolutely hated *Walden.*

"High brow *tripe*," he said. "What anybody sees in a book about a godless tree-humping bookworm living out in the woods in a lopsided old cabin is quite *beyond* me."

"It's about the value of self-sustenance and the innate divinity of nature and man," I argued.

"What in the *hell's* so all-fired divine about Thoreau?" growled Fuke. "When that crazy killin' fool John Brown tried to put Virginia to the torch, Thoreau was *against* his execution. Son of a bitch was all for *draggin'* Southern men and women out of their beds and cuttin' 'em up, like a painted *savage.*"

"Just the slave owners," I said, thinking of my father's many sermons to me on the rightness of abolitionism.

"Slave owners have as much a right to live as anybody," Fuke said. "Any man who says otherwise doesn't have a *shred* of divinity."

"Did you ever own slaves?"

Fuke laughed.

"Slaves were a *luxury*, and we didn't have many of those. Ask Mr. Thoreau what he knew about luxury. Only a rich *fool* would choose to go live in the wildwood if he didn't have to."

"Have you ever read *Walden?*" I nearly shrieked.

Fuke shrugged.

"I have not, nor do I *care* to," he admitted. "'Lord, my heart is not *haughty*, nor mine eyes *lofty*: nor do I exercise myself in great matters, or in things too *high* for me,'" Fuke recited, looking at me. "You could take a lesson from that, *Juniper*. You and all those high hats."

I retorted: "'When the dust of the earth has choked a great man's voice, the words he said will turn oracles.'"

The contest might have continued, except Jack interrupted us with:

"'Bird don't sang jest to hear itself.'"

Fuke and I both turned to regard the Missourian with new interest.

"Who said that, *Fats?*" Fuke asked.

"My paw, oncet," Jack answered.

From ahead of us, War Bag's laugh rumbled out like a rolling barrel and he called over his shoulder without turning.

"Fatty, your paw's got a passel more sense than these two! Yes sir, a passel!"

We forded the river at a shallow point around two 'o clock, and by half past three had reached the 'Picket Wire,' or Purgatoire River. Situated on a bluff on the north bank, was the old stone bastion of Fort Lyon, surrounded by a small camp of Indian tepees.

We crossed the river.

Sixth

Lyon consisted of a few buildings and a wide parade ground encircled by sixteen foot stone walls. The stars and stripes swelled on a pole. Red legged infantrymen were drilling when we entered, invisible below the waist in the cloud of dust from their stamping feet. A burly German sergeant cursed them up and down.

We were hailed by a young buck private at the gate. He directed us to the adjutant's office, and suggested we stop over and give our names and our business in the Territory. On the far end of the yard was the place in question, a low log building where all the freighters and Indians passed to make their marks or sign their names.

The inside was militarily clean, with a handsome wood banister separating us visitors from the officer and his clerks. Beside the little swinging gate of the partition stood a turkey necked corporal with cropped hair and freckles in a baggy uniform. Behind a big desk sat a gruff and gray flecked Major with bottle green eyes and a cigar about the girth of a tree trunk clamped between his hard jowls.

The corporal stood at attention, but his eyes were far away. They came into focus and looked us over. His attention piqued a bit at Fat Jack's unusual size before fading back into apathy.

A pair of Indians turned to leave as we came in. The corporal opened the little gate for the Indians and we all parted to let them pass. I caught their scent in my nostrils as they passed. It was a spicy, unwashed smell, not entirely unlike Jack. Different, somehow.

The Major wrote in his book and looked up at us briefly before returning to his work.

The corporal held the gate open and said, "If you'll proceed two at a time . . . ."

War Bag and Roam went in first. The protocol was ridiculous. The bannister provided no privacy, for the Major had a stern voice, and I don't believe War Bag had the ability to whisper.

"Gentlemen," said the Major.

"'Lo, suh," said Roam.

War Bag nodded and said, "Major."

I moved about to stay out of the line of the corporal's vacant gaze. I had the irrational city-bred fear of looking into another man's eyes, even by accident.

"Can you both write?" asked the Major.

I watched over the corporal's shoulder as Roam and War Bag nodded and the Major offered them the big ledger which they stooped over the desk to sign.

When they had finished, the Major turned the book back around and looked it over.

"'Ephron Tyler'," he read, "'out of Indiana. Bound for Kansas.'" The Major looked up at War Bag. "'To hunt buffalo.' These men are all in your outfit?"

War Bag nodded.

The Major leaned back in his chair.

"You know, there was a time when we had to fire the artillery pieces to keep the buffalo from scratching their flanks on the walls. You'll find the pickings a little slim these days, I'm sorry to say. Kansas isn't much better."

Behind me, Fuke whispered to Jack,

"I *told* him."

War Bag said, "We'll do alright, I guess."

"Have you enough lead?"

"I planned on petitionin' the sutler for some," War Bag admitted.

The Major nodded.

"You'll find him accommodating. You know, most of the hunters are Texas-bound now. The herds are all heading south."

At that, Roam said, "That's against treaty, suh."

The Major looked at Roam sharply, seeming to notice him for the first time. He looked down at the ledger, and read aloud.

"'Roam Welty.'" He looked up at Roam again. "Well, Mister Welty. Where'd you get that tunic you're sporting?"

"Tenth Cavalry, Company H, suh." said Roam, suddenly rigid.

The Major's expression eased some, and he took the big cigar out of his mouth, tapping the ashes into a brass tray with one carrot finger.

"Company H. That's Louis Carpenter's command, isn't it?"

"Yassuh," said Roam.

"Are you a deserter?"

Roam's eyes flashed briefly.

"Nossuh."

"Not that you'd tell me if you were."

Roam looked at the Major.

"I mustered out at Ft. Sill four years ago, suh."

The Major narrowed his eyes at Roam, considering his tunic again.

"Were you at Beecher's Island?"

Roam paused.

"Yassuh," he said.

War Bag cleared his throat. It was like gravel rumbling in an alligator's gut.

The Major shifted his eyes to War Bag, then back to Roam. He placed the cigar back in his jaws.

"I served with George Forsyth in the War," he said. "I guess he owes the Tenth Cavalry a lot."

Roam looked slightly embarrassed.

"Well suh, I guess we was just followin' orders."

The Major nodded.

"I guess."

"How about Indians, Major?" War Bag interrupted.

The Major shrugged.

"Oh, I don't think they'll bother you, Mr. Tyler. Not in Kansas." He looked slyly at them both. "Of course in Texas, the Kiowas and the Comanches have been a bit agitated as of late."

War Bag nodded.

"We'll stay out of Texas."

Roam said nothing.

The Major smiled, tapping ashes into his tray again.

"Of course. That'll be all. But mark my words, Mr. Tyler. You'll make no profit in Kansas. There's a bill before the Kansas legislature right now to regulate buffalo hunting, which essentially means to ban it. They're even lobbying here in the Territory. Some people say we're killing too many." The Major raised his eyebrows. "But there are none left to kill up here at any rate. It makes sense to go where the game is good. Any old hunter knows that."

War Bag turned around and passed through the little gate. He looked grim. He passed between us and stepped outside without a word.

"Thank you, suh," said Roam.

Roam turned and followed War Bag out the door.

Being an odd numbered bunch, Jack, Fuke, and I all went to the Major together, and the corporal did not object.

When we had all signed our names (save for Jack, who made a mark), the Major read it over.

"'G.P.F. Latouche, buffalo runner, Baton Rouge, Lousiana.'"

"*East* Baton Rogue, suh," said Fuke.

The Major ignored him, and looked up at Jack, raising his eyebrows at the big Missourian.

"And what's your name?"

"Jonathan McDade," Jack said quietly, watching the Major pencil in his name.

"And from where do you hail, Mr. McDade? Mount Olympus?"

"*Brown* Mountain," said Jack. "In Mad'son County, Missoura."

Squinting, the Major read my name.

"That's me," I admitted.

"From . . . ," and he paused, frowning at the paper. "Is it Chicago?"

I nodded.

"It says you're out here 'collecting material,'" said the Major, looking up at me. "What sort of material?"

I was a little embarrassed, for I thought perhaps he thought I was being deliberately cryptic.

"For a book, sir. I'm a writer."

The Major raised his eyebrows.

"Well," he said. "I'd work on my penmanship if I were you, son."

I had been ribbed by family and friends for my chicken scratchings all my life.

"Forgive me," I said, thinking of the old excuse I had learned from my mother. "I'm a left-hander."

The Major turned the page of the ledger, not looking at me.

"So am I," he said.

While War Bag and Roam went out among the freighters with their big wagons, Fat Jack, Fuke, and I went into the sutler's store.

Some Indians were squatting around the sutler's post, mostly a ragged, sorry looking bunch. They were not the majestic, noble people of the wild I had expected, nor even the quiet, strong type such as I had encountered in the person of Caddo Charlie. These were Utes and Cheyennes who had submitted to white authority, Fuke told me, and would lend their daughters' virtues out to any man with a jug of whiskey. Some were wolfers or bone pickers, who roamed the plains selling off the leavings of the white hunters to the railroads, who shipped them back east for soap. Others were destitute beggars, waiting to be shipped off forcibly to the reservations.

A shingle outside read 'Matthew J. McKintry, Licensed Post Sutler, U.S.A.'

The sutler's store was a wood and stone affair, well provisioned. Stacks of blankets adorned the shelves, and crockery and all manner of apparel hung from racks on the far wall. A great white buffalo skull looked down on the store from above the door, staring mournfully with empty eyes.

On the counter was a set of steelyards and a sizable pile of wolf pelts. There was a row of knives under glass. Presiding over it all was Matthew J. McKintry, a slim man with a big moustache. He was arguing with a shaggy Indian with a black crow feather in the band of his hat.

"Look, there's just not much of a market for wolf hide right now," the sutler said, in a sympathetic voice.

"When will you buy, then?" said the Indian.

"Why don't you take to picking bones? The railroad's buying them up like crazy."

The Indian shook his head.

"Got no wagon."

"Well, hire a teamster," said McKintry. "There's plenty here in between jobs."

"How'm I to hire a wagon without money from wolf skins?"

McKintry looked at the Indian quietly, then fingered the pile of hides, pursed his lips, and shrugged.

"I'll give you three dollars for these, but this is the last time."

The Indian nodded eagerly.

McKintry reached behind the counter and came back with a few coins.

"Don't use this for whiskey," he warned.

The Indian took the money from him and went out the door.

The sutler watched him for a moment, then sighed and dumped the pelts to the floor behind the counter.

"Hey *Juniper*," said Fuke. He and Jack were standing near a locked chicken wire cabinet.

I came over. Inside the cabinet were a few feathery Indian artifacts, including a kind of rattle made out of a turtle's shell, a feathered bonnet, and a hatchet.

Fuke pressed his finger against the glass and directed my attention to a beaded pistol scabbard with a decorative metal conch that caught the light, and a few strands of streaming fringe.

"It's boilt deerskin, I reckon," said Jack.

I did not for a moment think that an Indian had made that pistol scabbard, and wondered about the authenticity of some of the other items in the case.

"Juniper, I think that might fit that old *cannon* of your's," said Fuke. "Maybe you should get it before you have an *accident*." He looked at the Volcanic, protruding from the waistband of my pants pointedly.

"With what money?" I said.

"Maybe you can *trade* him something," said Fuke.

We left Jack in the store and went out to forage among Stillman's saddlebags and the parfleche Caddo Charlie had traded me for anything of worth.

After rummaging through my pack, we decided the nonessential items most worth trading were the blanket, a keen edged hatchet, and the books. I was reluctant to include *The Confidence Man* in the pile I carried into the store, but Fuke assured me Melville was a hack and I wouldn't have leisure to read it anyway.

We found the sutler and Jack engaged in a barter session. The Missourian looked obstinate.

"What would you trade for that beaded holster in the cabinet, *sir?*" said Fuke.

The sutler answered immediately.

"I'd trade for that bent knife," he said, eyeing the big arched blade on Jack's belt. "I've never seen the like."

"Let him have it, Fats," Fuke said. "What'll you *do* with a bent knife?"

The knife was a strange thing. It was oversized and heavy, with a cutting edge that bent sharply forward, not back. There were three notches etched into the base of the blade, just above the shaped, thick wood handle.

Jack would not have it, and backed away from the counter, as though we might try to take the Hindoo knife from him.

"Well, Ah might need it yet," he said, fingering the steel-capped handle.

Fuke shrugged.

"*Jingo!* Like a kid with a piece of hard rock *candy* . . . go watch the horses you big oaf," Fuke growled.

The big man sulked and went outside.

"We're short on cash money, but we've got some *items* to trade," said Fuke, motioning for me to lay out my wares.

The sutler was already shaking his head.

"I don't need another blanket," he said.

I laid the pile down on the counter, and the sutler rifled through it, his eyes dim, until he lighted on *The Confidence Man*. I should have expected it.

"I'll let you have it for this book, the hatchet, and the blanket," the sutler said.

"I thought you said you didn't need a blanket," I said.

"You can't expect me to give you a good scabbard for a beat up old book. I've got to even the trade somehow."

"It's *Melville*," I said.

"You can't wrap your ass in Melville come January," said the sutler.

"He has a *point*," said Fuke.

"What about the hatchet, Melville, and Shakespeare?" I offered.

The sutler shook his head.

"I've got Shakespeare coming out of my ears. My wife, she thinks she's bringing us all culture by reading to the troops and the Indians on Friday nights. Have you ever sat with a bunch of Indians and listened to Sonnets, with a cup of punch on one knee, and a dish of *petit fores* on the other? At least the troopers make a pretense of staying awake." He held up the copy of *The Confidence Man*. "I'm not exactly an admirer of Melville, but right now anything's preferable to Shakespeare."

I felt like striking this little tumbleweed literary critic for besmirching Herman Melville, but Fuke cut me off.

"*Alright* then, we'll take it." Fuke said.

Sliding that scabbard onto my belt and then fastening it around my waist for the first time was a sensation which stayed with me throughout my life. These days, wearing a gun is steadily passing from the inexplicably conjoined whim of fashion and legality. But strapping on that gaudy rig and feeling it hug my hip, it was as though I had completed a step towards some grand initiation, and had received a badge of sorts to merit my accomplishment. In my mind I had shed another layer of soft skin in favor of a bold new mold. A rush of excitement flooded through me, as I imagined the picture I made in my plainsman clothes with my beaded gun belt and unshaven face (for I had decided to cultivate a mustache and perhaps chin whiskers).

Roam shook his head at my new accouterments when he came into the sutler's.

"That shore is a fine scabbard you got there," he said, smiling.

I had elected to keep my pistol the way I had seen him wear his. He noticed.

"If'n you goin' wear it that way though, best to cinch it tighter. It's got to be up close, so's you kin grab hold of it whilst ridin.' Wearin' it butt first like that makes it safe t'draw while holdin' on the reins. You got to cock it and twist as you swing it up, see?" He mimed the motion for me, and I nodded understanding.

Fuke snorted and shook his head at some joke that the rest of us couldn't appreciate.

"We got us a muleskinner," Roam said, and passed me a handful of dollars. "Buy us a tub of arsenic, salt, hunnerd pounds of feed, fifty of corn, and a couple pounds of coffee. If'n I think of anythin' else, I'll come and say. We'll swing the wagon about front."

"What's the arsenic for?" I asked, as Roam went back outside.

"The coffee," Fuke said. "Gives it *bite*."

But he grinned lopsidedly, and I knew this time he was greening me.

Our muleskinner was Monday Loman. He'd been a farmer in Kansas and had lost his crop to the grasshoppers. Selling off most of his possessions, he'd left a wife and infant son back in Haskell County and taken an odd numbered seven mule team (for six he said, was the Devil's number) and a big Studebaker full of lamp oil and horse feed to Ft. Garland. He'd been running freight for nearly every military outpost and railroad camp between Ft. Sedgewick and Ft. Lyon since February, trying to raise the cash money to sink into the replenishment of his land. Greed and impatience had gotten the better of him recently, and he'd lost a good deal of his eggs in a 'sure-fire venture' that a pair of easy money men had sold him on in La Junta. He was desperate to make the money back and return home to his family in time for Thanksgiving.

Roam had convinced him that hauling hide and meat for our outfit was the way to do it, and assured him that he'd have enough money by the end of the season to get home and stack the dinner table for the holidays.

By now it was well known that War Bag held a particular distaste for grangers. Still, the wagon was a great incentive and additionally, the man claimed some knowledge of cooking. It seemed that War Bag did not have a fondness for mules either, for I heard him grumble about having to buy corn for the pesky animals, saying,

"Any beast that ain't got the sense to eat grass oughta be rubbed off the face of the earth."

The muleskinner had large, sad eyes that seemed in a constant state of peaceful reflection. He was a bit older than Fuke, with crooked teeth and blonde, stringy hair. His face was smooth, and the space between his eyebrows pock scarred. His chest was sunken and his arms thin. It seemed that had one of us punched him in the gullet, his extremities would have popped right out of their sockets. His long fingered hands hung like tear drops from his skinny wrists. Both ring fingers were adorned—one with a plain wedding band, the other with a Freemason's signet. His handshake was peaceful but firm and the palms of his hands were like leather. He spoke barely above a whisper, and Fuke seemed to take a certain delight in loudly demanding him to repeat himself.

Monday Loman's team of mules were as ornery and stubborn a bunch of animals as had ever walked the earth. He had named them after the archangels, and I believe those lofty titles had instilled in them a haughtiness not found in common examples of the breed. Yet the humble Monday somehow got these insufferable beasts under his control. Though they moved neither for curse, whip, or kick from any stranger, they readily submitted under the gentle touch of his reins and his low, quiet encouragement. Perhaps as archangels, they saw their benevolent master as the Supreme Being. I cannot tell.

While Fat Jack and I loaded up the wagon with the provisions Roam had ordered, a young lieutenant approached us and cordially invited us to attend a recital of Shakespeare at the mess hall at seven o'clock.

Bemused, I realized it was Friday, and Mrs. Matthew J. McKintry would be giving the recital. There would be punch and *petit fores.* I was curious to see the frontier counterpart of the stuffy tea parties and readings I had been forced to attend in Chicago, and resolved to go.

Fuke agreed to accompany me, expressing a fondness for Shakespeare, but War Bag and Roam showed no interest. After much talk, we did manage to persuade Fat Jack and our newest companion, the muleskinner, to accompany us.

"*What?*" Fuke said, inclining his ear toward the new man with delight.

"I said, I don't know that I'll understand it, but I'll attend," Monday Loman said again, barely louder than before.

"You kin set by me, then," said Jack.

Fuke reached up, clapped Jack on his burly shoulder and laughed.

"Fats here thought *Shakes-Spear* was a Comanche war chief!"

Fuke and I laughed. Monday Loman smiled thinly, and Jack's face flushed. He took off his hat, turned it over in his hands, and put it back on.

"Let's go," said Fuke.

War Bag said, "Don't go gettin' roostered up. We're pullin' out early."

"I never drink," the muleskinner assured him.

Seventh

The Thespian who would be our entertainment for the evening had not yet arrived when we strolled into the mess hall at a quarter to seven.

The long tables and benches were pushed to one side and the space filled with orderly rows of chairs. Soldiers were given seats near the front (enlisted men behind the officers) facing the mock stage area, and the independent traders and Indians occupied the back of the hall. One of the spare tables had been covered with a clean checkered cloth and lined with tin plates (McKintry china for the officers) and ironware (McKintry silverware for the officers), and the German sergeant we had seen earlier brought out a big wooden bowl of red punch and set it there with a ladle. There were paper napkins (cloth for the officers) and two big pans of cornbread (frosted *petit fores* for the officers). A pair of privates in aprons appeared with knives to dish out the food.

"I shore hope they don't cut that cornbread," I heard Fat Jack mutter beside me.

"Why, Jack?" I asked.

"*Bad* luck," he said.

The officers all rose suddenly, and I saw the sutler and his wife walk in, escorted by the adjutant and his missus, and a foppishly dressed man with a big stomach and oiled hair.

Mrs. Matthew J. McKintry was as proud and as cultured a lady as any I had seen in Chicago. She glided across the room with the poise of an aristocrat, one dainty gloved hand resting on Mr. McKintry's arm. He looked like a man on his last march.

Mrs. McKintry was a woman in her latter thirties. Her hair was in a tight bun from which sprung the occasional unruly black strand. She had the appearance of a woman struggling to maintain her fine breeding in this uncouth wilderness of blue wool and buckskin. Her face was generously dappled with paints and she carried a frilly parasol under one arm, though it was dusk and there was no need.

Behind the sutler and his wife came the Major, accompanied by one of the most beautiful young women I had ever seen on his arm. She was a fairy of a girl, with big eyes and two exciting rivulets of auburn hair that framed a clear white face. She had a slightly wide mouth with full lips, ears like apricot-halves, and a dainty chin that begged to be guided by gentle thumb and forefinger to the lips of a man.

She had none of the pomp of her elder counterpart, nor a need for it. I saw all the other faces of the men in the room turn to her, and the officers bowed in unison.

"Good e-vening Miss Pen-rose," they all said at once, like schoolboys.

In front of me, a toothless old bullwhacker in a leather coat who had the face of a baleen whale when he grinned, half turned in his seat and leered at me gummily,

"Some filly, hey boys?"

We nodded our dumb agreement.

Lastly came a foppish, pot bellied man in a long-tailed crushed velvet coat and ruffled shirt. The thin hair above his fat face was shoe-black and shined in the light. It had been slicked and styled so that it curled over his ears. He had a thin waxed moustache, checkered pants, and shiny little shoes. He walked with a melodramatic stride which I knew belonged only to one class of people; actors.

Sure enough, as the Major and Miss Penrose and the sutler took their prearranged seats in front of the officers, Mrs. Matthew J. McKintry beckoned for the man to stand beside her, and waited patiently but quite obviously for the soldiers, traders, and Indians to cease their muttering.

The fancy man stood beside her, chest puffed out like a strutting pigeon, nose upturned, oily pomade glistening in the lamp light. His coat was an affecting royal blue. All around, the men were snickering and whispering, but he seemed oblivious. The bullwhacker turned around again and said to us with a wink,

"Who in hell's this pork-eater?"

Mrs. McKintry at last raised one hand for silence, and the stern look she shot the young troopers got them all to dry up. The two spoons at the food table were the last to stop snickering, but the big German cleared his throat like a disapproving father and they too fell silent.

"Gentlemen," said Mrs. McKintry, in a flowery voice, "and . . . Miss Penrose." She nodded to the vision of womanhood sitting next to the post commander. I saw the two cut ups at the food table nudge each other.

"I have a special treat for you all this evening. Some of you may already know that we have had a distinguished guest in camp for a few days now. For those of you oblivious to the more important goings on, I am very pleased to introduce to you the esteemed Mr. Willard T. Waynesboro, Esquire."

With a flourish, she indicated the plump man in the

velvet coat. He bowed, smiling, his eyes half-lidded with the expectation of praise. When it didn't come, I felt a little sorry for him. But he just straightened out and smiled toothily, as though he'd just received a standing ovation.

"Mr. Waynesboro," Mrs. McKintry continued, her 'r's rolling off her tongue like lumber from a runaway wagon, "is an accomplished stage actor from New York City, and has performed many worthy roles," and she turned to her guest for confirmation. "Haven't you, Mr. Waynesboro?"

"Yes indeed, Madame," he said loudly, his big red mouth moving in an exaggerated manner around every word. He swivelled his great torso slightly left and right as he spoke, as though addressing every man in turn, but his eyes were focused somewhere behind and above us, so he succeeded only in gaining the full attention of the lath above the door. "I have enacted Napoleon Bonaparte, Polonious, (and with that, Mrs. McKintry smiled delightedly and turned to the troopers, as if to say 'you know Polonious,' though they showed no recognition) Henry the Fifth, and Richard the Third."

The old bullwhacker turned around again and said:

"He looks t'me like a blue davenport in a Kansas City whorehouse."

"I have performed before the King of Spain and the Queen of Scots," continued the actor, with much bravado, "but I dare say I have never played a venue quite so . . . lively."

Mrs. Matthew J. McKintry squealed and clapped her hands like a giddy little girl. I craned my neck and saw her husband put his head down. He was covertly leafing through *The Confidence Man.*

"I have asked Mr. Waynesboro to read with us tonight from the Bard, a selection of his choosing," she said to the troopers.

No one breathed, but the post commander clapped lightly and the officers applauded, followed by the enlisted men, some of the Indians, and Fuke.

"This should be *good*," he said.

Waynesboro bowed again, and when the smattering of applause had died down, he smiled pleasantly.

"I have chosen to perform *Titus Andronicus*," he said.

The men muttered to each other, some shaking their heads.

"Have you read *that?*" Fuke asked me carefully.

I shook my head. I hadn't.

Fuke folded his arms.

"It's a *good* one," he said, nodding and clapping, though I got the feeling he hadn't either.

"I oncet knowed a feller called Titus what lived in a cave in Taney County," Jack whispered. "He was a water witch."

"Shut up, *Fats*," hissed Fuke.

But I was intrigued.

"What in the world is a water witch, Jack?"

"A water witch is jest a feller what's got the power to witch underground springs and sich. Oncet my paw needed a well dug and we called on 'Ol Titus. He come right over with a witch stick about yay long (and he fenced off a space of about three feet between his big hands) cut from a peach tree. He stepped 'round the land like this, (and he made like he was holding two somethings angled downward) and shut his eyes." Jack shut his eyes for effect, before continuing. "And when he come to a certain spot . . . ."

But Fuke shushed him then, as did the bullwhacker and two Indians to our right.

"Before we commence tonight's recital," Mrs. McKintry began, "you can all line up to receive . . . ."

Before she finished her sentence the Indians and traders got to their feet, as did some of the troopers, until the officers turned around and fixed them with discouraging looks.

". . . punch and cornbread." Mrs. McKintry glared at those that were already getting in line (us included), and then smiled to the post commander and the officers. "Ladies and officers first, please."

Some of the troopers groaned and sat back down as the officers and the ethereal Miss Penrose slowly stood up. The rest of us milled around in the back, the Indians pacing and watching as the stiff officers helped themselves to pastries and punch.

Then the regulars got their turn, and we slowly filed in behind. By the time half the men were returning to their seats with precariously balanced plates and saucers, the punch bowl was half drained and the cornbread nearly gone.

"Aww," sighed Fat Jack, when we finally neared the table and he saw the cornbread cut up into neat sections. "It's s'pilet." His shoulders sagged and without another word, he turned and went back to his seat.

I had taken Stillman Cruther's Shakespeare book out of Caddo Charlie's parfleche and was thumbing through *Titus Andronicus* when a small hand touched my shoulder and I found myself looking into the decorative face of Mrs. McKintry herself.

"What's that you're reading, young man?" she said, smiling at me.

I was a little embarrassed, feeling the eyes of the other freighters and my own companions turn on me.

I held up the cover of the book for her to read.

"Shakespeare, ma'am," I said quietly.

"Well, well!" she cawed shrilly, and waved over Mr. Waynesboro. "Mr. Waynesboro, do come see this."

Mr. Waynesboro shuffled over. A bit of white icing roosted on the side of his face like bird droppings.

"It appears we have a budding thespian right here in our very midst!" said Mrs. McKintry, a little more excited than was warranted, I think.

Mr. Waynesboro looked me up and down and took the battered book from me without leave, reading the cover.

"Well, well!" he said. "Tell me, my boy. Do you read this?"

Then I got a little annoyed. What in hell would I be doing with a book otherwise? Wiping my backside?

"I've read it some," I said, trying to retain my dignity and yet not wanting to appear too lofty to my cohorts.

"Have you read *Titus Andronicus?*" he asked.

"No," I admitted.

"Ah!" he said, rather pleased, handing the book back to me with a fresh smudge of chocolate on the binding. "It is the very essence of tragedy. All the elements of Shakespearian drama. Passion! Betrayal! Skullduggery! Proscenium!"

He looked at Mrs. McKintry. I could feel the refreshment line moving on behind me.

"Mrs. McKintry . . . ," he began.

"Yes, Mr. Waynesboro?" she replied, still smiling.

"Might we not find a place for our young hopeful in the recital?"

"Oh yes, let's!" she agreed emphatically. I think if he would have suggested I play Lady Macbeth and she a Moorish gravedigger all in blackface she would have been ecstatic.

Waynesboro produced a thick leather bound edition of Shakespeare, trimmed with gold leafing, and opened it up. He ran his finger lightly across the page and then turned it so I could see it. His pudgy finger, like an uncooked sausage, pointed to the words "YOUNG LUCIUS: *A boy. Son to* LUCIUS."

"It's perfect for you," he assured me.

Mrs. McKintry patted my shoulder.

"If you do a good reading, I'll introduce you to the Major and his ward afterwards," she told me.

That was quite enough to keep me from protesting.

"Thank you, ma'am. I'd like that indeed." I turned and shook the actor's hand. It was soft and fleshy, sticky with frosting. "Thank you, sir."

"I'll have a chair brought up," Waynesboro said. "Have some punch. You can sit beside me."

"See you soon," trilled Mrs. McKintry, and whirled away.

I turned back to the food table to find one of the privates scraping the last of the cornbread from the pan with a fork, and an Indian boy dunking a piece of frybread from his pocket into the punch bowl and sucking at it.

Fuke and Monday were standing next to the bullwhacker, talking when I walked up.

Fuke turned to regard me.

"Your socializing has cost you the common doings, *Juniper*," he said, munching the last bit of his cornbread pointedly.

I shrugged.

"Jack said it was bad luck anyway."

"What the *hell* does he know?" he said around a full mouth.

Monday held up his crumbling piece of cornbread.

"You can have this, if you've a mind." he said.

I shook my head.

"It's alright, thanks."

"That little wife of the Major's is sure one *crackerjack* lady," Fuke said, looking over my shoulder.

"That's not his wife," I said. "She's his ward."

"How do you know, *Juniper?*"

"Mrs. McKintry said she's going to introduce us after the reading . . . if I do a good job."

"You opportunistic *cuss,* whyn't you invite me to read?"

I shrugged.

"If you want, I will."

I guided Fuke over to Mrs. McKintry and Mr. Waynesboro, and he made a big show of his upbringing in East Baton Rogue and his love of Shakespeare, *particularly* . . . ah, *Titus Andronicus.*

Mrs. McKintry seemed much taken with him. More so than with me it seemed, for she transferred my promised introduction immediately to my more charming rival.

Mr. Waynesboro too was fascinated. It was hard not to be, I suppose, with the Lousianan's smooth talk, shining eyes, and singular dress. Soon he had been assigned the coveted spot alongside Mr. Waynesboro and the role of Marcus Andronicus, brother to the lead. I was demoted to sitting beside Mrs. McKintry.

The part of Lucius the elder was to be read by still another candidate for Miss Penrose's attentions, the same young lieutenant who had invited us to this gala affair earlier.

A Christian Indian with glasses and a silver cross on a chain called Jay Proud Dog was told to read the part of

Saturninus, the stage directions and the part of Aaron, a murderous Moor.

The remainder of the roles were split between Mr. Waynesboro (Titus Andronicus, and several supporting parts), Mrs. McKintry (Queen Tamora), the young Lieutenant (Lucius The Elder), and the angelic Miss Penrose herself (Lavinia, daughter of Titus).

When everybody had settled down and put their plates beneath their seats, we all sat down in a semicircle in front of them. Mr. Waynesboro passed out some worn reading copies of the play and gave Jay Proud Dog the signal to begin.

Jay read slowly and carefully, but well. As he read his opening lines, Mr. Waynesboro got a very serious, wide-eyed expression on his face and began to rise slowly and dramatically from his seat, even while Jay was reading. Jay kept glancing from the page to Mr. Waynesboro. Mr. Waynesboro's attention was somewhere above the door.

When Jay read the lines:

"Then let my father's honors live in me, Nor wrong my age with this indignity."

Mr. Waynesboro sprang up to his full height, his knees popped loudly, and proclaimed in a booming voice:

"Romans! Friends! Followers! Favorers of my right!"

And he gestured violently to his right, where Mrs. McKintry was sitting. She threw up her arms in surprise and nearly fell from her chair.

A couple of the troopers burst into laughter, but their officers silenced them.

So began my foray into frontier Shakespeare. Leadville got Oscar Wilde. In Ft. Lyon, we had Willard T. Waynesboro.

Eighth

Titus Andronicus went along as the evening wore on. Mr. Waynesboro hammed it up while the rest of us delivered our lines much as they were printed on the page, in dull, monotone voices.

Mrs. McKintry tried to mimic Mr. Waynesboro's flourish, occasionally leaving her seat and twirling her spindly fingers. Fuke read slowest amongst all of us. I was embarrassed for him. His ears grew hot whenever his tongue stumbled, which was often.

I never got a chance to read at all.

As the others read and I followed along, occasionally leaning over to whisper the pronunciation of a word in Fuke's ear, I began to skip ahead. In the course of this, I discovered a couple of things.

Firstly, Young Lucius (the part that Waynesboro had said was perfect for me) did not have a single line until the second scene of the Third Act.

Secondly, *Titus Andronicus* was the grisliest piece of Shakespeare I had ever read.

The play concerns a Roman general (Titus) who conquers a Goth queen (Tamora) and slays two of her sons, all for the honor of the Emperor Saturninus. As in most Shakespearian tragedies, there is a heap of revenge, betrayal and misunderstanding which leads to a bigger heap of bloodshed on either side. *But what bloodshed!*

Titus loses his hand, his sons are beheaded, and his daughter Lavinia is raped by Tamora's sons over the corpse of her husband. Then they cut out her tongue and lop off her hands so she can't betray them to Titus. Of course he finds out anyway. Understandably upset by this, Titus gets revenge on Tamora's sons by killing them, cooking them, and feeding them to their mother. Now to be sure, there is a lot of story in between all this which I have omitted due to spacial constraints, but it's my hope that the reader gets the gist of the mood of the play if not the drama.

I had read all this by the time my colleagues had reached the Second Act, and by then had begun to understand why I had never heard *Titus Andronicus* read at any of the ladies' socials I had attended in Chicago.

I looked at Mrs. McKintry, who had no idea that in another Act she would be dining on her own children, and

Miss Penrose, whose pristine beauty was about to be defiled and mutilated by Mr. Waynesboro. He gleefully read every line of the two Goth villains, knowing full well what was to come.

I wondered what sort of mind counted *Titus Andronicus* among their fondest Shakespearian reads, and what sort of ghoul would recite it with a pair of ladies. I wondered if, in his tactlessness, Mr. Waynesboro would seize upon the opportunity to try and act out the coming violence upon poor Miss Penrose, as he no doubt had done for the Queen of Scots before being belted out of that country. It was no wonder he was confining his performances to Colorado.

I looked at the audience. Most of the officers were rapt, as courtesy demanded. A good deal of the troopers were awake, though fighting to be so. In the back, I could see Fat Jack and Monday the muleskinner watching and listening attentively. Behind them, some of the Indians were sleeping, most noticeably an old man in a corner who was snoring to beat the band.

In particular though, I noticed the Indian with the crow feather I had seen hawking wolf skins in McKintry's earlier. He was smiling and bleary eyed, and passing something back and forth between himself and an equally happy companion.

"Give me thy poinard ('pwa-nard,' was what she said)!" Mrs. McKintry exclaimed, the rapture in her voice sorely misplaced. *"You shall know, my boys. Your mother's hand shall right your mother's wrong."*

"Stay, madame!" read Mr. Waynesboro, rising up. He made a dramatic sweep with his arm, as if restraining her, and took a step toward Miss Penrose, who by now had gotten used to his behavior and ignored him. He leered at her then, and said without glancing at the book, as though he knew the words by heart,

"Here is more belongs to her. First thrash the corn. Then after burn the straw . . . drag hence her husband to some secret hole, and make his dead trunk pillow to our lust."

At the word *lust,* most of the men in the audience perked up. Fat Jack and Monday grew wide-eyed, and the two Indians looked at each other. Everywhere men jostled their companions awake. In the front, the Major's expression sharpened.

Mrs. McKintry, I think, was only waiting for the chance to speak, and not really paying attention to the words. I saw Miss Penrose blink a few times. She no doubt had begun reading her dialogue ahead. The Lieutenant cleared his throat and looked sheepishly at the Major in the front row.

Mr. Waynesboro did not even pause in his rendition.

"Come my brother!" he raged, turning to a brother who was not there, but speaking instead to a distracted looking

Fuke. *"Now perforce we will enjoy that well-preserved honesty of hers!"*

The Major straightened up in his chair at that. Beside him, McKintry leaned forward eagerly, and *The Confidence Man* slid discarded to the seat beside him.

"O, do not learn her wrath, she taught it thee!" pleaded Miss Penrose (she was apt). *"The milk thou suck'dst from her did turn to marble; Even at thy teat thou hadst thy tyranny!"*

At the word "*teat*" Fuke scratched his head and his face turned beet red.

The men were all awake now. The two at the food table were open mouthed.

"What begg'st thou, then? Fond woman, let me go!" Mrs. McKintry yodeled.

"Tis present death I beg!" said Miss Penrose. *"One that womanhood denies my tongue to tell! O, keep me from their worse than killing lust, and tumble me into some loathsome pit where never man's eye may behold my body! Do this, and be a charitable murderer."*

"*So should I rob my sweet sons of their fee?*" shouted Mrs. McKintry, standing up and sitting down quickly, so as not to lose her place on the page. *"No, let them stay and satisfy their . . . lust (*her voice dying in her throat*) . . . . on . . . . thee?"*

"No grace?" whimpered Miss Penrose, getting into the spirit and doing a very convincing job, by the looks of the men in the troop (though not quite in the way the Bard had intended). *"No womanhood?"* she said quietly, her soft face fallen, her eyebrows arched, her doe eyes suddenly inclined to regard Mr. Waynesboro. *"Beastly creature! The blot and enemy to our general name! Confusion, fall"*

"Nay! Then I'll stop your mouth—" began Mr. Waynesboro, but he never finished the line, for Fuke's fist collided with the side of his face, knocking him right to the floor.

A general commotion went up among the crowd. The troopers began laughing and whistling. The two Indians stood up in the back and whooped excitedly. One of them had a bottle of whiskey in his hand, and he raised it in salute. Fat Jack was visibly shocked, and Monday shook his head.

Beside me, a deafening double shriek pealed out of Mrs. McKintry and Miss Penrose both. Jay Proud Dog the Christian Indian knocked over his chair to get out of the way. The Lieutenant sprang to his feet, but did nothing.

Fuke stood over the fallen Mr. Willard T. Waynesboro, Esquire and balled his fists.

"Get up, ya sumbitch, and I'll stop *yore* filthy mouth for ya!" he shouted. All pretense of the southern gentleman was gone. He growled in the same sort of backwoods drawl I might have expected from an angered Fat Jack. I thought of him struggling with the words of *Titus Andronicus* and I was suddenly deeply sorry for him.

I looked down at poor Mr. Willard T. Waynesboro. His velvet coat had split the seams and his fat back was bursting through. He was sprawled on his large belly, rubbing the side of his swelling face with one pudgy hand. His meticulously oiled hair was a disheveled mess.

The officers got to their feet, unsure of what to do. The two aprons at the food table were flat out laughing.

The Major went swiftly to the side of the fallen actor and stooped to help him up.

I stood up and took Fuke by the elbow.

The Major looked up at us as the Lieutenant stepped over and stooped to help Waynesboro.

"You'd better get going," he said, not unkindly.

"Come on," I said to Fuke. I led him to the back of the hall, through the gauntlet of laughing troopers and Indians and past where Jack and Monday were hurriedly pushing through the audience to follow us.

Fuke was still red in the face when we got out into the cool night air. War Bag and the old bullwhacker were sitting out on the porch rail. They stood.

Fuke jerked his arm away from me and tried to shove past War Bag, but the old man was immovable, and caught his wrist in a strong grip.

Fuke hid his face, and turned away.

War Bag released him, but Fuke did not move.

War Bag put his hand on Fuke's shoulder and walked off the porch with him.

"No shame, son," I heard him say.

Jack and Monday joined me shortly on the porch.

"Is he alright?" asked Monday.

Jack took a step off the porch after Fuke and War Bag, but the bullwhacker restrained him.

"Let 'em be, big feller," said the bullwhacker. "He'll be awright."

Jack shrugged, watching War Bag and Fuke cross the parade ground. They were talking, but what they said could not be heard.

"What happened?" said Roam, emerging from the dark, a cigarette drooping in his mouth.

"Nothing," I said.

I watched the old man and Fuke walk off into the darkness and disappear.

I did not see them again until morning.

Ninth

Summer Came In The Morning.
We rode out of Lyon right into the beginning of a bright May. The sun drove over our shoulders jauntily as a honeymoon coachman.

War Bag had purchased a big Murphy wagon and twelve oxen team from McKintry. Jack volunteered to drive the rig, claiming experience in handling oxen. It was a good, sturdy wagon, with a steel bed and nine inch tread flat iron tires. War Bag said we would use it to haul hides. We had become quite an impressive and well-provisioned little army, and were intent on waging war against the buffalo.

Yellow orioles flitted on perches of long grass, and we saw geese returning to their summer haunts from their winter havens, forking across the sky like Indian arrows.

Monday Loman's supply laden wagon rumbled along. The farmer sung low hymns to entertain his mules, who brayed appreciatively now and again like a chapel full of Baptists. The odd mule was used for packing, and followed the wagon on a tether.

I found that in taking inventory of the angels I had ever heard named, I could count only four -Raphael, Michael, Gabriel, and Lucifer (but I wasn't sure if Lucifer was still associated with his more respectable brethren).

Monday Loman however, had much more than a passing knowledge of angels.

"See those two up there? I always put them in the lead. Not because I favor 'em particularly, but because they're more apt to pay attention to the trail. That one on the left, our left, is Uriel. He's named after the angel that warned Noah of the Flood. He's got a weather sense about him, and gets real fidgety when the air changes. That one on the right is Old Gabriel, the messenger. Let's me know about whatever Uriel don't catch on to. Behind him is Raphael. He's real calm and sensible, like the angel that whispers man's prayers into the Lord's ear. Next to him is Ramiel. He always sleeps later than the others. Ramiel is the angel that brings the dead from their graves on Judgement Day, so that's sort of a joke."

They all looked the same to me, but of course, Monday had spent more time socializing with them than I had.

"Now behind Ramiel, that's Raguel," he went on. "He

will bite you if you whip him. Raguel is the angel of vengeance that struck down the first born of Egypt. I always put Saraqael next to him if I can. He's got an obedient heart, and nips the others when they do wrong. Saraqael is the angel that pleads for those with sins of the spirit."

"What about the one in the back?" I asked, jerking a thumb at the pack mule that followed the wagon. He brayed and shook his bristly mane, as though he knew I was addressing him.

"He's the orneriest of the lot -likes to kick a lot. I named him Michael, after the one that kicked Lucifer out of Heaven."

By midday, the infectious harmony of the natural world around us had dispelled whatever remained of any post-drunk misery. It was good to feel the warm sun on the backs of our necks.

I thought to ask Fuke about the night before, or make some light comment about his one-sided bout of fisticuffs with Willard T. Waynesboro, but thought better of it. At any rate the chance never presented itself. Fuke spent most of his time riding beside War Bag, and spoke little.

"What's the word with Fuke?" I asked Roam.

Roam shrugged.

"He keeps talkin' in the ole man's ear 'bout goin' south to the panhandle."

"That bothers you?"

"Lotta runners headed south."

"Well, shouldn't we go where the buffalo are?" I asked, remembering what the Major had said, for it seemed logical.

"It ain't so easy. Comanche and Kiowas comin' off the reservations don't want us movin' in on their huntin' grounds."

"Aren't there plenty of buffalo to go around?" I had read somewhere that the buffalo traveled in tremendous numbers.

"Well," admitted Roam, "but white huntin' is different from Indian huntin.' Indian'll cut from a herd of buff 'til he reckons he's got enough, then pack up and go home. He'll make his food, his lodges, his clothes, even his arms all out of what he got offa that hunt. Most white hunters'll kill and kill till there ain't nothin' left, and take just whatever they figure they can get the most cash money for. They leave the rest to rot. For awhile it's been meat and tongues. Next year it might be just hides. Ever'body tryin' to glut the market, and meanwhile the Indians is losin' they commissary."

Roam shook his head.

"Well, I guess there'll be plenty of hunting to be done in Kansas anyhow, right?"

"Maybe," said Roam.

As we rode on, another thought occurred to me.

"What was Beecher's Island?"

Roam shrugged.

"Just a Indian fight."

"You were there?"

"You heard the Major," he said.

"So who's George Forsyth then?" I pressed.

"Forsyth was a officer 'got into trouble with some Cheyennes at Beecher's Island. The 10th was sent to relieve him, that's all."

"What happened?" I pressed.

Roam looked at me, and his eyes said much, though his lips said only,

"We done it."

He eased his pinto up beside Fuke and War Bag, leaving me behind with the wagons.

We saw no buffalo that first day, but Fuke shot an elk and we had meat that night around the fire. We pitched our tents in a fine spot with some rare trees to keep out any storm that might come, and a little creek trickling nearby.

Jack showed me the best way to skin the elk. He demonstrated the purpose of each of Stillman's cutting tools, and in seemingly no time he had stripped away the tawny hide of the beast and neatly separated its rack from its skull. He presented the antlers to Monday, who mounted them on his wagon.

"All'ays keep yer blades good 'n sharp, " he admonished, producing a butcher's steel which he scraped his knife across, honing its edge to a keen. "Y'got t'ile 'em, too. Keep 'em real clean."

His big hands were bloody as a murderer's from the work.

"You use the same knives to skin a buffalo?" I asked.

Jack nodded, thrusting the steel through his belt.

"I'm partial t'pullin' it off, but ye kin cut 'em 'round the hey-ed, hook the hide t'some chain, 'en wrap that 'bout the axle of a wagon. Then ye kin let the mules er a good hoss do most o' the work. Don' worry 'bout carvin' out the meat. I'll be doin' most of that."

I must've looked bewildered, for he said,

"Don't fret. I'll larn ye."

In the time that I been out in the wilderness, I had never seen a sky so clear at night. Perhaps it was the company, or the situation I was in that added to the effect. I was doing what boys back in the city dreamed of. I was sitting in a frontier camp, swapping stories and dining on freshly killed meat beneath a western sky. And what a sky! The stars and planets shone like silver dust strewn over rich velvet, and craning my neck I felt as though I were suspended over a bowl set to catch the diamond crumbs of a jeweler's work.

Monday cut the antelope into tender steaks, salting what was not cooked and packing it away in the wagon. He fried beans in a Dutch oven over the fire, and they sizzled and gave off an aroma that I had never known could be so sweet.

We dug in, and I think if Fat Jack could have collapsed his jaw like a rattlesnake, he would have induced Fuke's elk to leap antlers and all down his throat. He ate like a man with doom knocking.

As Jack reached for his third helping, Fuke said,

"*Damn,* Fats! I'll wager if the Lord hadn't put bones around your belly, you'd eat up every elk in *Creation!*"

Monday snorted as he passed Jack another shank of meat, and fouled the portion with an expulsion from his nostrils.

Jack looked at the meat sadly while the rest of us guffawed.

"I'm sorry, Jack," Monday said, mightily embarrassed. He ran his sleeve across his nose.

"My momma always said summer colds were the *worst*," Fuke commented.

Monday sniffled in agreement.

"Aw," said Jack. "It's just one side."

He set to cutting the snotted portion away.

Monday got up and walked out of the firelight toward the mules, sniffling the whole way.

After that display no one was inclined to eat any further except Jack, who gladly finished Roam's when he offered it. It seemed an eternity before he finally sat back, tucked four of his fingers into the lip of his drawers, and announced his surrender with a shuddering burp.

Monday returned to the fireside. Staring idly into the fire while we picked our teeth and dwelt on our meal, he began to sing "Bringin In The Sheaves" in an almost angelic voice.

War Bag joined in. Jack pulled out his Jew's harp from the string around his neck where he kept it, and in a minute a regular revival had commenced. I marveled at the rich baritone of War Bag's singing voice. I remember looking up and thinking that the ears of every illuminated head in heaven were inclining down to our campsite, fooled into thinking that between our muleskinner's boyish voice and War Bag's opposing baritone, the boy Christ and His venerable Father were engaged in a duet.

Then Roam began to sing a Negro hymn. I counted our outfit blessed that three such voices could exist and beyond all hope converge on our campsite that first lovely summer night. Roam didn't have the inhuman depth of War Bag, nor the elfin weightlessness of Monday, but it struck an even ground somewhere between, and rose and fell in a way that I don't believe any white man's voice could.

The song was called "Steal Away ," and went something like:

> *Steal A-way.*
> *Steal A-way*
> *Steal A-way to Je-sus.*
> *Steal A-way.*
> *Steal A-way home.*
> *I ain't got long to stay here.*
> *My Lord calls me,*
> *He calls me by the thun-der.*
> *The trum-pet sounds with-in my soul,*
> *I ain't got long to stay here.*

When his voice gave way, he stared into the depths of the fire, and the light played on his large dark eyes, dancing. I wondered at his thoughts.

"Don't *none* of you know anything but *church* songs?" Fuke said after a bit.

"I thought you were a Christian yourself, Fuke," I said, thinking of his testimony that reading the Bible was the best argument for literacy in the first place.

"I've nothing against loving Jesus, *Juniper*," he said, and I warmed, knowing he was himself again. "But there's such a thing as praying *too much*."

Monday looked upset by this.

"A man ought to honor the Lord every minute of the day."

"That wouldn't leave much time for else, would it?" Fuke said. "This here reminds me of the story about the riverboat captain that struck a *snag* near the Cairo shoals." He clasped his hands behind his head and propped himself against his saddle.

"As the ferry started its *descent* to the bottom, and the passengers were all up to their *ankles* in water, the captain come down from the texas and announced, 'Is there ah . . . any man among you who can *pray*?"

We all leaned in.

"So one little *minister* from ah . . . . *Cincinnati* adjusts his spectacles and raises his hand and says . . . ."

Fuke raised one hand, and spoke in a mock Eastern voice. Deprived of his usual Southern drawl and assuming a stiff posture, the end result was pretty hilarious in its own right.

"'Ah . . . excuse me. *Ahem.* I can, sir.'"

In an instant he was back to himself, beaming devilishly beneath his mustache.

"So the Captain says, 'Good. You start praying while the *rest* of the passengers put on the life preservers. We happen to be *one short.*'"

We all snickered appreciatively, except for Monday, who shook his head.

War Bag said: "Amen."

Tenth

The First Buffalo I saw was a stinking mound of decomposing flesh and fly infested hide.

Roam found it just before noon on the third day out of Ft. Lyon. He was riding ahead, and came across wolf sign in the earth. He followed it to a cracked and dry gully, whereupon he sighted a murder of crows circling in the air. Beneath them were two wolves up to their muzzles in the side of a dead buffalo bull, and two others playing at tug of war with the tongue.

Roam ended their conflict with two well-placed flights from his bow that left as many wolf carcasses lying in a heap. The two that had been feeding, he said, flattened their ears and bounded off at the quick across the prairie.

When the rest of us came up, Roam was retrieving his arrows from the necks of the fallen wolves and wiping the blood on the grass.

The buffalo bull was lying halfway down the slope of the gully, as though it had died in mid descent. Its great bearded chin was flat on the cracked earth. Flies buzzed in and out of its empty eye sockets. Beside the torn holes made by the wolves in its flanks, there was a great gash in its belly, and its intestines were strung out behind it like doll stuffings.

The bull was impressive even in death. I could see beneath the thick mane evidence of the broad muscles that propelled it across the land like a shaggy locomotive. But its horns were blunted and its coat flecked with mange, worthless as a hide.

"Old timer," Roam observed.

War Bag craned his neck, peering at the wound in the bull's belly.

"Been hooked. Probably the younger bulls kicked it out of the herd. I reckon it must have dragged its guts awhile."

"Is that normal?" I asked.

War Bag said,

"Young bulls always oust the old timers. No place for 'em."

"If we follow his trail back, we ought to find the rest of the herd," Roam suggested.

We followed the blood flecked path of the bull into Kansas, leaving Monday and Jack behind with the wagons to follow at their own pace.

When at last we found the great trampled swath where the buffalo herd had passed, Roam and War Bag were elated.

"They's headin' south," Roam said.

"Toward the *panhandle*, I bet," Fuke sighed, in a voice meant to remind us that he had said this all along.

War Bag looked at Fuke, and Fuke straightened up in his saddle and looked away.

"They ain't all gonna make it," War Bag said. "You go on back and make sure Fatty and that muleskinner don't get lost."

Fuke started to protest, but War Bag cut him off.

"You got the fastest horse, and Stretch here won't know how to follow us."

Fuke wheeled his horse about and went off in the direction we'd come, holding onto his hat as Napoleon broke into a run.

The rest of us turned into the wake of the buffalo. The trail was easy to follow, as the buffalo beat the earth down wherever they went. It was, as Fuke had said, a straight line south for the panhandle.

"Come next year, ain't gonna be no buff up here at all," Roam said.

"Next year," War Bag said.

After an hour we stopped to graze the horses and wipe them down. We had to steer our mounts a mile to the west to find good grass that the herd had not rendered inedible. I was tired of smelling buffalo dung, and said so.

War Bag said:

"You'll smell a lot worse than that."

"Trail's gettin' fresher. Must be they smell the North Fork

by now. Maybe the front of the herd's already there, and they slowin' up the stragglers," said Roam, sipping from his canteen.

War Bag looked to the north from whence we'd come.

"That mule team better get here. I don't reckon I'll shoot the whole herd myself."

He looked at me, and I suddenly felt as though he blamed my inexperience for the absence of his star shooter.

Roam elected to go on ahead alone and scout the herd, leaving War Bag and myself behind.

War Bag smoked while I chewed pemmican and jotted down notes about the country and thoughts I'd been having concerning Stillman Cruthers, Fuke, and Beecher's Island (which I could not find on any of Roam's maps).

I was obsessed with finding out little details of War Bag's past—especially concerning the cicatrice above his eye. I do not fully understand the fascination men have with divining the cause of bodily injuries.

I do not know if this morbid curiosity is shared by the reader, but I have often found myself looking with inordinate attention on the malformed—the legless, the armless, the blind, the scarred. I find myself concocting explanations in my mind for their condition.

In the case of those who have lost limbs, I always felt a certain special pity towards them. I remember seeing a man with a silver hook for a hand at the First National Bank and thinking how I could not fathom the feeling of losing a hand. How does one cope with the loss of such a constant companion? How mourn that which in youth brought toys close, and penned letters to loved ones in later years, or rested in the palm of a sweetheart? To have that ripped or torn away, never to return . . . .

War Bag drew his Henry rifle from its saddle scabbard and shot a skunk in the grass nearby, kicking me out of my musings.

"Damn thing," he said, touching the little carcass with the toe of his boot and puffing his pipe.

I remembered Roam's warning about skunks on the day I'd met him.

"Why kill it?" I asked, looking down at the small, broken form. It was glass eyed and a little pitiable.

"Not too long ago they were spreadin' hydrophobia ever'where. Bit the camp dogs, camp dogs bit the men."

We settled back down, but I couldn't concentrate on writing anymore.

"Roam told me Stillman Cruthers was a schoolteacher."

War Bag cracked a smile and fingered the scar above his eye absent mindedly.

"Still weren't no schoolteacher. He was a damn lawyer."

"How'd he end up buffalo hunting?"

"He shot another law dog up in Leavenworth."

"Killed him?"

War Bag shook his head.

"Still wasn't that lucky. Blew off three of the fella's toes. Right there in the courthouse." He shook a bit with laughter, his eyes shining with recollection. "Told me he was aimin' for 'the center of mass.' If he'd a'been a better shot, maybe he would've been alright."

"If he'd have killed him?"

"Sure. As it was, that lawyer man turned out to have a congressman for a cousin. That put Still on damn near everybody's list."

"Why did he shoot him?"

War Bag shrugged.

"Still was a queer sort. He read too much to be a criminal and he drank too much to be a lawyer, and he was too goddamned lazy to be anything else. He had this idea he was gonna make a name for himself one day, but it never happened. Just one of them fellas weren't meant to live long, I guess."

In some way I suddenly felt as though I were walking in his footsteps.

"Where did you meet him? Where did he come from?"

"Dodge City in '69. I think he was a Square Head."

"How did you meet?"

War Bag hardened, and his face grew stern.

"You ask too many damn questions, boy."

He picked up his Henry and began to clean it, and we spoke no more.

About a half an hour passed in silence and tobacco smoke before Fuke, Jack, and Monday came into sight from the north.

By the time they were in shouting distance, we had saddled up and mounted.

Fuke trotted up to us.

"Roam's gone ahead," War Bag said. "We got about a half hour's ride to the North Fork, where the herd'll be waterin' by now. We'll make camp on the south bank. It's a big herd. I 'spect we'll be well off for a week or two at least."

War Bag pointed Solomon south and we all fell in with him, trotting along, the mule team right behind.

*   *   *

I spied in the distance a congregation of live buffalo trotting south. They were big shaggy bulls, dark and woolly, their black horns traced white by the noonday sun overhead.

I gave an excited yell and pointed.

"It's just the old bulls trailin' the herd," War Bag said. "See how mangey they are, even for the weather? Ain't hardly even worth shootin.'"

But Fuke drew his repeater anyway and kicked his horse. Letting out a whoop, he took off after the small herd of twelve.

Some of the bulls craned their necks at Fuke's galloping approach and snorted. They broke into a run, and cut as a body sharply southwest, running for all they were worth.

The brim of Fuke's big hat turned up in the wind and he twisted his Winchester up to his cheek as he had done with the rabbit. He let out two shots that sounded like a pan banging on the roof of a tin shed in the distance.

War Bag muttered,

"Damn fool."

Fuke dexterously steered Napoleon toward the old bulls, weaving quickly left and then right, stinging them like a maddened wasp. Smoke spat from the barrel, and one of the lagging bulls stumbled and fell heels over head to the earth.

Another bull reeled drunkenly from the line, slowed, and stopped. It shook its great head, as if it were trying to rid its ears of a noisome ringing. Fuke spotted the wounded buffalo, turned away from the chase, and trotted nonchalantly up to within thirty feet of the old animal. It sensed him and charged, but Fuke jerked Napoleon safely away, circled, and watched as the bull ran for a few feet. It stopped and commenced to shaking its head again.

Fuke closed the distance and the bull wheeled and made for him as before, lowering its head like a cow catcher.

Fuke easily avoided it, and circled.

"Why don't he jest kill it?" Jack whined.

War Bag said:

"'Cause he's young and mean."

Fuke dodged the old bull twice more. Each time the animal's response was a little slower. Finally he rode right up to it, lowered his rifle, and levered three shots into it up close.

The bull sagged and collapsed where it stood.

Fuke walked Napoleon around the bull, leaning from his saddle to admire his handiwork. He trotted back to our party, smiling.

"That sure gets the *blood* up," he said, when he was close.

"Shall I fetch them tongues?" Jack asked War Bag.

"Leave 'em be," he said. "They ain't worth nothin.'"

I watched the old herd slow to the southwest and then continue on south as though nothing had happened.

"Do they just die?" I asked War Bag. "The old ones, I mean."

"They got no place anyhow. Makes room for the spikes—the young bulls."

I watched the bulls and thought of the carcasses we'd left behind. They would lay until the enzymes of the prairie came to break them down, the wolves and the crows and whatever else. Perhaps some scavenger would choke on Fuke's lead, and then it too would die and be consumed. But the bullets, like the prairie, would remain. The snows and the rains would corrode them, but they would yet lay there, more useless than stones.

We sighted the North Fork of the Canadian River. It was a vein of silver on the hilly horizon. What I had mistaken for hills from a distance was actually hundreds of dead and skinned buffalo, their bald bodies pink and white and caked with drying blood and clouds of flies.

The smell grew thick, and we all tightened our bandannas over our faces to filter the putrid stench. A few of Monday's animals balked at passing through the plain of death, but dutiful Saraqael kept them in line.

Jack's poor plough horse Foggy went berserk and bucked and crow hopped behind the hide wagon, till it was feared he would hurt himself or spook the mule team. War Bag wet a spare rag with his canteen and instructed Jack to tie it over the horse's muzzle. He did, and Jack was able to walk Foggy through the sea of spoiling meat.

All around us the buffalo lay, cow and bull alike, naked and humiliated in the sun. Furred masks were all that remained on their great heads, which lay at unnatural angles in the blood soaked grass. Their humps had been carved from their backs and carried off, and their rear legs were spindly and skeletal, the hams cut away. Their jaws were open, but where there should have lolled the precious tongues, there were cracked rivulets of black, drying blood and torn stumps. They were murdered victims left silent as Lavinia. They could not even plead for revenge in the ears of their buffalo god.

On the south bank, Roam waited on his horse. He took off his hat and waved to us, then pointed down river. Far in the distance were a wagon and a team of oxen, and a few figures bustling about still more dead buffalo.

We forded the river with a good deal of difficulty, for the land here was wet and sandy and every bit of help was required. Monday told his mules the story of Jesus and Peter walking on the Sea of Galilee.

More than a few times we had to dismount and bodily free the wheels of the wagons from the muddy earth, leaning against them and gripping the spokes like tenacious helmsmen in the midst of a vicious gale. We cursed profusely despite Monday's recitation of the Scripture. I felt as though I were blaspheming in the midst of a sermon, like the man with the unclean spirit forced to sit in a pew, the devils in him loudly voicing their indignation.

The waters of the North Fork were cool and insistent, tugging at my legs. The muddy banks sucked hungrily at our feet, trying to strip us of our boots. War Bag and Roam fixed ropes to the wagon and mounted on the south bank, spurred their horses to help pull them. After an hour's dreadful, back-straining toil, we extracted them from the water.

We sat there on the south bank for some time, the air sharp in our lungs, the mules and the oxen shaking the water from their flanks. Jack had a smallish cat with him.

He was a haggard young thing, with a good sized bite in its left ear and three scrawny legs with white socks. One leg had been lost to whatever prairie denizen had attempted to gnaw its ear off -probably a wolf or a raccoon. He was smoky gray with a hint of striping on its hindquarters, and big green eyes.

The cat clung to Jack's big belly and shivered. He meowed constantly and shook his wet fur, miserable.

"Where'd you find him, Jack?" I asked.

"'Bout three miles back. He come and poked his head outta the grass whilst Monday an' me was rollin' by, so I jest scooped him up an' took him with."

"Don't he never shut up?" War Bag moaned.

"You should call him *Pisser,* with the way he's pissing and moaning," Fuke said.

"Before, he hardly said anythin' at all. I named 'im Whisper," Jack said, stroking the cat's scrawny neck with his thumb and forefinger.

"He look like a drowned rat," Roam observed.

"He's jest chilled, s'all," Jack said.

"Let's drop him in the coffee pot and *warm* him up," Fuke suggested.

Jack pulled the little cat closer. He got his tack and

canteen from the wagon, and wetting a bit of pemmican, he cut it into little pieces and began to feed Whisper.

Roam groaned. He stood up and took off his hat, wiping his brow and gazing east toward the outfit of skinners down river.

"Somebody ought to go and see what they about," he said.

Though a good deal of the herd remained alive, it was courtesy not to try and weasel in on another group's find, just as it was for prospectors and herders. To ignore that rule was to incite hard feelings or even bloodshed.

War Bag told Fuke and myself to come with him, and we would go to the skinner's camp while Jack, Monday and Roam made sure the wagon was alright.

By the time we had ridden downstream, the warm sun had dried us completely.

Eleventh

The Skinners' Camp was a temporary affair erected to serve as a crude factory to process the dead buffalo. What hides had been gleaned already were pegged and stretched out on the ground hair side down for about a hundred yards. The skins were pale and yellow as they dried in the sun, while the beefy tongues hung from makeshift wooden racks. A smoke house was up and running, just a rickety thing of plank and twigs. The flavorful smell was pleasing, not unlike the burning of leaves, or a cooking flame.

Two men were sitting at a low table eating and drinking coffee as we rode up.

"Afternoon," called War Bag from the edge of their camp.

One of the men raised his tin mug and waved us in. It winked reassuringly in the sun.

War Bag dismounted and led Solomon the rest of the way. Fuke and I followed suit, and tethered our horses to their wagon.

A razor strop of a fellow with a wide mouth and matted blonde hair stood up and nodded to us.

"How do, friends?" he said.

"I'm Eph Tyler," War Bag said. Then he introduced Fuke and I in turn.

The other man put out one of his long hands and War Bag took it and gave it a friendly jerk.

"Dex Kellerman," said the angular fellow, taking his hand back and offering it to Fuke and I.

He motioned to his companion, a man with a walrus mustache and a fur cap who stood up at our introductions.

"This here's 'Zeck. He don't talk much American, but he's alright."

The man with the mustache shook our hands firmly and motioned to the coffeepot.

"*Fahrizeya*," he said, in a thick East European accent. I was not sure if it was a greeting in his language, or what. "Please. You dreenk."

War Bag accepted a cup of coffee, pouring it himself. We all waited like aides to a diplomat and then aped him. The coffee was hot and had a kick of liquor in it.

"Meanin' no offense, but is this your outfit?" War Bag asked Dex Kellerman.

"No sir, we're part of Harley Wade and Garvey Bledsoe's bunch, but I *am* top skinner 'round here."

"You from Kansas?"

"Mostly. We're headed down into Texas like 'ever'body else."

War Bag rubbed his chin and produced his pipe.

From the corner of my eye, I saw Fuke look over at War Bag expectantly.

"Ah . . . how is the hunting down in *Texas*?" Fuke asked.

"We hear tell they run in the millions," said Kellerman.

"Lot of runners headed south?" War Bag said, packing his pipe.

"Near every one," Dex confirmed. "Charlie Rath and Wright Mooar's set up shop at the old Adobe Walls. Griffin's gettin' to be a good place to sell hide, too. They're goin' for near a dollar and a half in Dallas, and you can unload buff hams for three cents a pound." He craned his neck upriver.

"You might've brought another wagon."

"You headed for the Walls?" asked War Bag.

Dex shook his head.

"We're gonna load up here and freight the meat on up to Dodge City. But we'll be back by the fall for hides, most likely winter at the Walls. Either that, or Rath City."

"What about Indians?" I asked, for I was anxious to see some in the wild.

A third man, filthy and covered with grime and blood, sauntered into camp from the east. His clothes were dark and wet with gore, and the sharp knife in his belt was stained red and strung with viscera. He looked like a newborn, fresh from the womb.

"God damn all Indians," muttered the bloody man, as he

stepped up and poured himself a mug of coffee. He brushed against me, and his touch left a red smear on my sleeve.

"This's Pete Adderly," Dex said, winking at us.

The man stopped sipping his coffee and turned to regard us.

"I'll tell you about Indians," Pete Adderly growled, from beneath his ruddy mask of filth and cruor. "I'd ruther get shot in the belly by a white man and dumpt in a shaller ditch than fall into the clutches of one o' them murderin' red savages. *Yessir.*"

"Why is that?" I asked.

Pete Adderly looked at me.

"You get caught by them and scalpin's the least of your worries. A Indian'll skin you alive and cut out your tongue—leave you lyin' in the sun jest like a goddamned stinkin' buffalo. *Yessir.*"

He spat in the dust.

"If you're smart, you'll get yerself one of these."

He tugged at a bit of twine around his thick neck, and produced a small oblong object from where it hung beneath his shirt.

I peered at the bloody thing, and saw that it was an empty rifle cartridge.

"What is that?" I asked.

He set his coffee down on the table, took the cartridge in both fingers, and popped it open. Inside was a smaller cartridge, and inside that, a tiny glass tube of white powder.

He leered at me, and his teeth were as black and dirty as the rest of him.

"Hydrocyanic powder," said Pete Adderly. "Bite the white. Kill ya quicker'n a bullet in the brainpan. *Yessir.*"

"*Poison?*" I exclaimed.

He popped his deadly vial back together and let it fall back next to his heart.

"Indian won't scalp a dead man," he told me. "Leastways, y'won't feel it if they do."

Then, as if his own harsh talk had upset him, he slung what remained of his coffee to the earth and strolled back to the killing field.

"God damn all Indians," he muttered in parting.

Everyone was quiet for a minute, and it seemed to me Dex was a bit embarrassed by his comrade's tirade.

"Pete's old outfit got massacred by Cheyenne," Dex explained. "As long as you keep a couple of good needle guns at the ready, you should be alright." Then, as an afterthought, he said, "If you like, you could probably ask to join up with our outfit. Safer that way."

War Bag shook his head.

"Thanks, but I believe we'll head south."

"Well, alright then," said Dex. "Good luck to you."

When we returned from the skinners' camp, a debate ensued over what our next course of action should be.

Fuke reiterated his desire to drive into Texas, where we had been reminded again and again that the hunting was best. Apparently in his hour's ride with Monday and Fat Jack he had convinced them the merits of this plan as well.

"If the money is to be made there, then that's where we should be," was Monday's reasoning.

"We ain't seen naught but rotted carcasses, I reckin," Jack said. "I 'spect if them men say Texas is where the buffalo is, we ought to go."

Roam was still against it however.

"Indians goin' see any man south of the Arkansas as bustin' treaty," he said, "and they'll be lookin' to raise hair."

With Pete Adderly's poison vial still in my memory, I was convinced of the clarity of Roam's judgement, and said so.

That left our leader sitting on the fence.

He put his boot up on a spoke of the hide wagon wheel and knocked the leavings out of his pipe on his boot. He put it away and gazed over the land, where the skinners scurried like busy ants.

He folded his arms, and it seemed that by looking south, he was peering into all the possibilities of the future, observing and weighing the outcomes of every course of action. Like an oracle, he had already seen it all transpire a thousand different ways, a thousand times over. Weighed down with that knowledge, could he prevent any ill consequences, or was that the limit of his power? Was he an eye with no hands?

War Bag said,

"We'll go see about Texas."

And though it was warm, a chill gripped me briefly, for I believe now I caught a harrowing glimpse of what he had seen and forgotten.

Twelfth

Bullthrower saw action for the first time a week later.

The going got rocky, and we had to lead the horses bodily across the rough terrain. We were all tired, and had been looking forward to making camp along the river.

It had been an uneventful week, with sparse game and no buffalo or men in sight.

Jack let his cat down from the wagon and it crept along beside us, weaving in and out of the islands of mesquite brambles, chasing the occasional field mouse, or pausing to groom itself on one of the numerous flat rocks.

Whisper grew quickly from a scrawny, skittish stray to a staunch member of our party. Only four days ago he had been clinging to his fat messiah and would not come near another soul. Now he seemed perfectly comfortable with the lot of us, save for Fuke and War Bag, who kicked or cussed at him whenever he came near.

I wondered how he had gotten all the way out in the middle of Kansas prairie. In my mind, I wove all manner of stories for him. Mostly they entailed who his owners had been, how he had lost his leg, the battles he must have fought, the near escapes, and the long nights spent cowering beneath the brush as some feline-hungry horror stalked about, snuffling in the dark.

He ran like a whole cat, bounding through the grass as though his missing leg were invisible to our eyes. The hip joint moved in sync with the rest of his body, flexing muscles to animate a nonexistent leg.

Whisper ran the mice down for the sheer thrill of it, as Fuke had run down the buffalo bulls. He toyed with them. When they were caught too quickly, he let them go to satisfy the lust that must have been born in the beating hearts of his predatory ancestors, when the cat was a monster that bore sharp fangs three feet long. To his credit, when his prey would or could run no more, they were devoured wholly, fuel for the fire.

Whisper, as Fuke had observed, was not at all a fitting name for the cat. He was quiet only when left alone. When seeking affection he meowed incessantly, and would not quiet down to hear the Final Judgement.

More than anything this fed the hatred of War Bag and Fuke, who complained as long as Whisper crooned. The cat

(Fuke continued to refer to him as 'Pisser') was threatened with all manner of bodily harm, from a protracted 'goodnighting' of his reproductive organs with War Bag's hip knife, to a sound stomping from Fuke's high heeled boots.

Consequently, the cat was warned off from seeking the pair's attentions by the rest of us, and learned on its own not to incur their wrath.

At the Goff River, Whisper had just caught another field mouse when at last Roam called out what we had all been waiting to hear,

"Buffalo!"

While Monday's piety seemed to forbid him the use of firearms (he did not even carry a squirrel rifle, and when questioned about the wisdom of traversing the plains alone without any kind of protection, he would just shrug, as though that were answer enough), War Bag's horse Solomon carried a veritable armory.

There were times when I watched War Bag slouch in his saddle and pass his slitted eyes over the open country while his black horse dipped its nose in water or tall grass. At such times, silhouetted against the setting or rising sun, he reminded me of a mounted fighter of olden times, his rigging adorned with battle implements, his countenance scarred and beaten from some recent, bloody row. With his long plaited hair and beard, he looked like one of those savage mail-clad Vikings I had seen in picture books as a boy.

There were two rifle scabbards hanging at all times from either side of War Bag's horse. One of them was a faded, fringed buckskin type, and bore his Henry. The other, which he drew forth that day, was a stiff, enclosed case of cracked, hard leather.

We crossed the river and he and Fuke dismounted and unlimbered their long arms.

Fuke took out an eleven pound Remington .44. It was a mean looking apparatus with a long, serious barrel and thick wood stock. As a bludgeon it must have been formidable enough, but as a rifle it was sure death.

When War Bag undid the fastenings on his leather case, he drew out the largest, deadliest looking rifle I had ever seen.

It was huge, with a long, octagonal barrel, thick, polished wood stock and furnishings, and a pale, carved 20 power telescopic sight mounted on the top. The stock was reinforced with iron or steel plates, and nestled in the underbelly was a fancy double-trigger.

Roam must've seen my eyes widen at the sight of it.

"Man look like he ain't never seen a rifle before," he said.

War Bag smiled faintly as he worked.

"She's a Sharps," Roam said, with admiration in his voice. "Fourteen pounder. Shoot anythin' from a .44 to a .52 caliber, dependin' on the barrel."

"That's no *rifle,* it's a field piece," Fuke muttered, polishing his own weapon, as if it needed it.

I saw that the oil Fuke used came from an unmarked container, and asked him what kind it was.

"Eagle oil's the *only* thing I use," he replied.

"How much did that tin of axle grease cost you, George?" War Bag asked, chuckling. "And what drummer sold it to you? He ought to be ashamed of himself."

"Not likely, *old man,*" Fuke said stiffly. "I brought down and milked the old *buzzard* that went into this bottle myself." He mimicked taking aim with the Remington, at some invisible bird which I imagined diving majestically, sounding its death-scream as Fuke drew a bead. "Got him as he dove for a *jackrabbit.* Clean through both eyes."

War Bag just shook his head and rummaged through his rifle case. Beside the bullet molds, powder and makings, there were three extra barrels set securely in neat housings.

"What're the sights made out of?" I asked, for any I had seen on rifles had always been crafted of sparkling brass.

"Buffalo leg bone," War Bag said. "Cuts out the glare."

War Bag took a ready made cartridge out of his canvas flap belt and fit it into the breech. It was about three inches long, and I imagined it smashing through hide and flesh and bone with all the velocity of a locomotive. He sent it home, and the mechanical sound was not without authority.

"We call her Bullthrower," Roam smiled.

I looked at Fuke, who was silently hefting his rifle and mock-aiming at the buffalo on the low ridge Roam had sighted.

"What do you call that one?" I asked him.

Fuke shrugged, looking at his rifle.

"*Mine,*" he said.

The ten or twelve buffalo Roam had spotted were not old bulls, but young cows. They stood on a low ridge about a hundred yards out, grazing complacently as we made our preparations.

"I'm goin' scout the herd," Roam announced. He rode his horse over to the ridge.

At his approach the cows did not bolt, as I would have expected, but turned around in no great hurry and walked away.

While Roam was sizing up the crop, War Bag presented me with a small leather bound ledger.

"That's yours. You keep track of how many buff you skin each day, or when we divvy up, you won't get paid what you worked for. I don't expect you'll be bustin' any records today, but that's alright. You just stick by Fatty and he'll teach you all you got to know."

He turned to Monday then.

"Sin Buster, you stay behind with these two. Haul the meat and hides back to camp. If you wanna pitch in and help 'em, the payoff'll be bigger, and you'll have a better stake to take back to your little wife. Mind, you're responsible for keepin' your own tally. I trust you boys won't try to pull the wool over each other, and I know you won't on me."

To Jack, War Bag said: "You're the butcher and the boss skinner too, Fatty. I hate to have to saddle you with both these babes, but I believe you can handle it."

Jack seemed to color a little bit.

"Aw, I'll larn 'em awright, I reckin," he said.

War Bag turned and took his Henry rifle off his horse. He handed it to Jack. From his saddlebags he produced a box of cartridges.

"You keep this 'till you get one of your own."

Jack beamed with pride. Even though it was just an old Henry, it may well have come from the King of England.

Roam returned.

"They's pretty well spread out. Near a hunnerd an' fifty of 'em. Plenty a places t'shoot, too." He pointed over at a low jumble of stone. "One of you can probably get to within two hundred yards over on the west side. They's a little stand of rocks. You'll be upwind."

War Bag nodded, and walked over to the camp wagon. He took out a forked stick from the kindling pile. Taking Solomon by the reins, he set out for the east flank of the herd. Fuke did the same and headed off after him.

"You fellas goin' follow behind whilst we cut from the herd," Roam said. "Y'all skin any carcasses. I'll be ridin' back and forth betwixt y'all, to check on you, an' I'll help you spot 'em. Monday, you make camp over that ridge, once the herd moves. Be sure and gather up some chips for the fire. Dig a pit, too."

"Will you be back for lunch?" Monday asked.

Roam shook his head.

"You don't have to worry about nothin' but supper. I'll be by later, and we'll put up a smokehouse."

Roam mounted and rode off in the direction War Bag had gone. Before he loped out of sight, we heard the first booming reports of the buffalo guns like approaching thunder.

That first day Bullthrower left forty two buffalo scattered about on the prairie, and Fuke accounted for forty eight. A hundred buffalo were dead by evening, and we were skinning almost by torchlight.

Of the seventy or so hides we were able to successfully claim before nightfall, I marked twenty down in my ledger. At my promised rate of twenty five cents a hide, in one day's (admittedly hard) work, I had barely earned ten dollars.

I began to wonder if it was worth it.

Second Part: Buffalo Days

"Whilst skinning the damned old stinkers, our lives they had no show,
For Indians waited to pick us off on the range of the buffalo."
—from *The Buffalo Skinners*

First

Skinning Buffalo was no easy task. The abominable nature of my work accounted for my feelings of dissatisfaction that first evening, when I sleepily did the figuring in my cramped ledger by the light of the campfire. I had not felt so downright contemptible of work since the railroad.

Jack called the animals 'big stinkers.' I found this unique to the lexicon of the skinner, for no shooter had to deal with their quarry at this base level. As a beast of the field, the bison gave off the same natural odor as any other animal. But the stench that emitted from its carcass once the shaggy hide was peeled away was comparable to no other olfactory phenomenon I have ever experienced. It comes off the flesh in waves, and flushes the cheeks like a hot summer day, raising the bile and curling the lips. The skinner wears stench and parasites like a uniform.

The insects are as unavoidable as the blood and reek. They leap eagerly from the hide as you stoop to remove it, bouncing jauntily up your arms or burrowing deep into your clothing. The flies buzz in your ears and brush their wings against your lashes, and no bandanna will keep the fleas from seeking your nostrils. By the end of the day's work a skinner is a stinking, bloody mound of bug bites and flea colonies.

"Mind yer hands," Fat Jack would warn me always. "If'n ya get cut it'll hurt yer tally."

That first day my tally was the last thing I thought of.

Fat Jack helped me skin the first three buffalo we came across. He would swiftly carve the profitable portions of meat (the tongue, hump, and hams) from the bones and lay them out on the hide for Monday to pick up and load into the wagon. A greater amount was left on the skeleton for the buzzards.

Jack watched me do the next three, but after that it was sink or swim. Most of the time I floundered. I know for a fact that I was responsible for ruining at least seven or eight hides and mangling or splitting as many humps and tongues. Despite the filth, skinning was a skill that required a degree of finesse. A cut or torn hide was worth nothing, as was meat that looked as if it had been once through the grinder already.

Finally, in his meek way, Jack suggested I not try and butcher the animals. It meant double the work for him, but he did it without complaint.

So I set myself to concentrating on the perfection of my skinning technique.

The easiest way for me to skin one of the beasts was to enlist the aide of Othello. Monday used this method the few times he joined us in skinning, calling on one or two of his 'archangels' for help. He warned me that it was a sketchy practice, and not very easy on the hides.

Monday began the same way Fat Jack had taught me. He cut the buffalo beneath the jaw (the first time I did this I inserted my knife too deep and pierced the jugular like a grapefruit, spraying myself in the face with warm blood), and traced a line around the jaw and down the belly all the way to the tail. Next he ripped the tough hide from the carcass down the inside of each of the four legs, unwrapping the buffalo like a box at Christmas.

The skin about the back of the neck and the ears (leaving the wooly scalp and bearded face attached) was sliced by a sickle shaped knife and the hide gradually loosened from the carcass.

Then came the final step, where Monday's and Fat Jack's methods diverged. On the first three buffalos Jack and I skinned successfully, the big Missourian taught me to use the weight of the big animal to tear its own skin off. Jack made it look easy, but when left to my own, it proved a daunting task.

Monday showed me how to attach ropes to the skin at this stage and by way of a horse, peel it cleanly off. He would hook the rope or chain to the rear axle of his wagon and the other end to the carcass, inducing the mules to peel the hide in increments.

Having no team of mules, nor the strength of the Missourian, I chose the middle ground with Othello.

The chocolate gelding proved to sport a disposition ill-favored to buffalo skinning. Twice when I was sure I had the hide prepared just right, the awls ripped away, leaving ragged tears in the skin. Three times after that, Othello took it in his mischievous brain to bolt at my gentle encouragement, and tore or split the hide badly. At those times I would wind up chasing the horse with its whipping, dancing ropes taunting me all the way. Luckily for me he usually ran right to Jack or Monday, who were never out of sight.

We trailed the herd and set upon the buffalos left fallen in the wake like eager scavengers. The herd was often in sight, and the booming of the guns never left my ears.

They were magnificent looking creatures, but apparently not possessing of even the most basic instinct of self-preservation. I was amazed that they did not stampede at the sound of the guns. They were like some utopian society unfamiliar with the concept of death—at least, not with death as the white man dealt it. I had heard of buffalo breaking into a run at the sound of thunder, and the old bulls had fled at the sight of Fuke's horse. If they had learned to equate the rumble of thunder and the gallop of horses with danger, why didn't they recognize the report of the big fifties as the same?

After too many had fallen, the remainder of the herd would trot off, only to stop nearby. Then the grisly process would start all over again. There was no sport in it. The buffalo did not run or cut suddenly to avoid the killing. They did not interpose themselves between some semblance of cover (usually because there was none), unless by accident.

War Bag knelt on a low hill and fired until he was a shadowy phantom in a fog of powder smoke. He did not need his horse. He hunted from afoot. Why were these men called runners? They did no running. To my eye, they just reclined easily on the hill, shooting at leisure while I was up to my elbows in filth. Meanwhile the buffalo fell like dim children, milling about and lowing, not having souls to grieve or the sense to run.

After I had skinned ten buffalo and was thoroughly bitter, Monday came around to collect the hides and the meat which Jack carved from the bone. He announced that he was going to go and set up camp and start staking the hides until Roam came to set up the smokehouse.

Roam did not visit us as promised, and I saw little of him as the herd moved out of sight. I did see him ride back to camp, but he did not return.

Night finally fell. When Monday came upon me struggling, he helped me with the last of the buffalo. He was kind enough to tell me I could mark them as my own and he would not contest it.

When we were finished, he told me he had been slow cooking a couple buffalo humps, and that we would have tongue as well, and muffins baked in his Dutch oven.

The hump was as tender and as succulent a meat as I have ever tasted. The tongue lived up to its much vaunted reputation, though having seen the dead buffalo up close and the things which congregated upon its palate, I was reluctant to try it.

An additional and unexpected treat came in a pile of boiled bones that were passed around and cracked for the dark marrow. We sucked or spread this on Monday's corn muffins as a salty butter. The flavor was like nothing I have tasted since.

War Bag, Fuke, and Roam sat in their own semicircle with the old man in the middle, and swapped stories about the day's shooting. I was resentful of them. Their job seemed easy beside all that I had endured that day.

The rest of us were too exhausted to speak. We filled our mouths and said nothing. Jack fed his cat scraps from his tin plate, and Whisper caterwauled after each bit.

"Fats, why don't you get *rid* of that damn cat?" Fuke hissed across the camp for what seemed the umpteenth time.

Something in this galled me. I felt like War Bag and Fuke were lording it over the rest of us, and their dislike for the cat and Jack's affection toward it somehow made the quarrel mine. Whisper was a skinner-cat, one of us.

"Why don't you shut your mouth, Fuke?" I growled.

The gravelly tone of my tortured voice must have startled them, for they all looked at me. I was bone-tired and irritable.

"What's in *your* soup, *Juniper?*" Fuke asked, in a manner that perturbed me.

"Nothing," I muttered, and chewed my meat, staring into the fire.

Twenty five cents a hide, was what I thought, again and again. *Twenty five cents!* Really it was a good deal of money. But all I knew was it would be months before I saw it.

There was not much more talk around the campfire that night, aside from the shooters. I heard War Bag speak of continuing south after the meat and hides were cured, for the band we had killed were only a part of a greater body.

I only vaguely understood his words. I am sure that I was asleep before my head touched my saddle.

After breakfast, Jack spied me itching and swatting at parasites. He walked me over to a side of the camp, and instructed me to lay my filthy clothes down on a good sized anthill that he pointed out in the dirt.

"When do we get paid?" I asked War Bag.

War Bag looked over at me. He was puffing his pipe.

"Why? What d'you intend?"

"I intend to cut out when we hit the next town," I told him firmly, loud enough for all to hear. I think at that moment I actually meant it.

This statement brought a ripple of low laughter from my comrades.

Fuke shook his head as though he had known what I would do all along.

"One day of *honest* work and you're ready to take the first train back to Chicago." He threw up his hands. "But what do you *expect* from a blue scissorbill fresh off his momma's teat?"

"That's an easy remark coming from you, you riverbottom ass," I shot back. "You don't do anything but talk. Any fool can sit on a hill all day with a rifle and kill a bunch of stupid animals. You wouldn't . . . ." I stopped myself. "You *couldn't* spend a day dipped in buffalo blood if your life depended on it."

"Juniper, you don't have an *inkling* of what the *hell* you're talking about," said Fuke.

"Why don't you prove it to him, Fuke?" said Roam.

He didn't say it in a way that said he doubted the Louisianan's words, but Fuke glared at Roam just the same.

"What're *you* saying?" he accused.

Roam shrugged.

War Bag said to me,

"You think any one of us ain't willing to do your work, boy?"

I was quiet. That was exactly what I thought, but I folded my arms and said nothing.

War Bag went up to my saddle. He reached down, plucked my tally ledger out of my gear, and opened it.

He sneered at the figures and let it fall to the ground.

"You're a charity, fella. Ain't worth throwin' on the kip pile."

I felt my ears color.

"I'd just like to see him skin one day, and I'll shoot and we'll see . . . ," I began, my voice cracking.

But War Bag cut me off with his voice, so heavy it was like an anvil falling.

"You don't do no shootin' till you prove you can earn your keep."

"And who decides that?" I snapped.

He stalked right up to me, and I took a step back. I thought he was going to hit me. Instead he stopped a foot in front of me and glared into my eyes.

"*I* do."

He looked over at Roam.

"You take Bullthrower and Fats'll tend you and George."

"Whatchoo goin' do?" Roam asked.

"I got to contend with that mangy cat complainin' 'cause Jack favors him," War Bag said. "But I ain't got no reason to put up with this scabby little nipper." He pointed a finger at me, and looked me in the eye. "I'm gonna make certain sure you're worth the trouble of keepin' around. And if you ain't, you can have your pay and my boot in your ass to see you to the next town."

Then he turned away, leaving me flushed and angry.

Roam went to his saddle, took out his Spencer rifle, and handed it over to War Bag, along with a cartridge belt, which the old man buckled on.

"We're goin' south after that herd," he announced to everyone as he drew the belt through the buckle and jerked it tight. Then he turned his eyes on me, and they were hard. "You keep up, boy."

I turned to lay my shirt down on the ant hill as Jack had instructed.

"Got no time for that," War Bag growled behind me. "You're so damn smart, you wear what you got."

In that moment I hated him.

Second

The Competition between War Bag and I began not more than an hour after we broke camp.

Roam spotted the remainder of the herd we had cut the previous day. He and Fuke and Jack rode off ahead, leaving me in an uncomfortable silence with War Bag that Monday tried to lighten with idle talk.

"It's a warm day, isn't it?" he said.

"Warm enough," War Bag replied.

"Back home I'd be praying for rain."

"'Rain follows the plow,'" War Bag remarked, quoting an advertisement I'd read somewhere.

Monday turned to War Bag, his face brightening.

"Have you ever worked the land, Mr. Tyler?"

"Some. Before I got sense."

The waiting was terrible. It was like sitting on a pregnant thunderhead, waiting for it to bottom out underneath you. War Bag gripped his pipe tightly in his teeth and said nothing more, but his face seemed to ripple with contained anger. He did not like me, and I was sorry for that.

I watched him tether Solomon to the back of the bull wagon and climb into the driver's seat. The smoke from his pipe drifted across his gnarled face.

I found the queer rock I had picked up along the banks

of the Rule, and ran my thumb across its coral-face in my pocket.

Monday staked out the hides all around our camp. They were stretched on the ground to dry in the sun, their underbellies skyward. Monday dashed them with the barrel of arsenic to kill the pests, walking between the rows as though he were dispensing powder to babes in a nursery.

Monday also tended the meat-vats. These were great holes dug in the earth and lined with pegged green hides. The hams and humps were salted just like beef or pork, placed in the vats with boiling brine, and covered tightly again with buff hides. After nine days of this treatment, the meat would be ready for the smokehouse.

The smokehouse was dug out of the side of a low hill. It would be lined with hides when complete. A low fire trench would be dug too, and the meat hung there to smoke.

"Let's go," War Bag said to me, rolling his sleeves when the popping of the guns began.

We came to within fifty yards of the fallen buffalo. There were eight by the time the herd began to move, clearing the way for our work.

War Bag got down from the bull wagon and went right to work without saying a word.

I rode off a ways and set to work on another carcass.

The going was hard as the sun was hot. There was the added irritation of knowing War Bag was in the corner of my eye taking close account of my progress. The vermin jumped with joy at my arrival, and the heat intensified the stink.

I ruined the first hide with a sloppy cut that left a shallow nick in my thumb.

I cursed aloud, but War Bag showed no interest. He hitched Solomon to the carcass of a bull and stripped the hide as Monday had done.

I hurried over to my second bull. This time I recalled the words of Jack, cautioning me to spare my hands. My knives were razor sharp, and I had no desire to test the efficiency of frontier medicine. I made my incisions carefully as a surgeon, and stood up to admire my work. It was undoubtedly the best cutting I had done thus far, and I was almost sad to have to trust the rest to Othello.

War Bag finished his first hide and moved on to the next without a glance in my direction. Maybe he wanted me to fail. Maybe he wanted me gone from the outfit.

The blood was sticky on my hands, and I fumbled with the awls and the ropes, but got the whole rig worked out as sturdy and as sure as I could. When I urged Othello on with a click of the tongue, I was delighted to see the skin come off clean a full foot.

I knelt down again and did as Monday had done, slowly helping the hide along and then returning to Othello, who continued to be surprisingly mild and cooperative.

I had the skin off in what seemed no time.

War Bag was on his third animal. He was butchering too.

Discouraged, I went back and stripped the hams, tongues, and humps from the carcass. I found I could butcher a buffalo with some skill.

The spur of competition was in me now, and I veritably hopped to the next carcass, a light colored cow that caught my eye. If things had gone well the last time, it was doubly so this time. The cow had a soft, malleable hide that was easy to cut and came off nicely.

War Bag was again hooking up the ropes to Solomon.

I moved on to another bull.

I do not know if War Bag's presence had some motivating effect, but I found myself cutting into that bull with all the dexterity of a practiced skinner. I was so pleased with myself I couldn't help but smile. When I got up to hook the ropes to Othello, I looked over at War Bag.

He was stooping over another beast, but had just turned his face away from me.

The work was tough, but invigorating. The sweat soaked my shirt and mingled with the buffalo blood. It dripped down my nose and I tasted the salt on my tongue. My ears were hot and my legs stuck to the insides of my jeans. My fingers slipped on the bloody skins and I caught myself with a handful of bare buffalo. I took the bandanna from around my neck and wrapped it around my knife to keep it steady in my hand. Immediately the sweat and blood ran down my back.

I had done four more buffalos when Jack rode up to War Bag. I did not concentrate fully on the conversation, as I noticed War Bag had not stopped skinning all the while Jack spoke. But the exchange ended with Jack exclaiming, "Why, ye've left yer canteen!"

War Bag said: "I guess it's in camp."

"Here, ye kin have one of mine," Jack said.

"No thanks," War Bag said. "You keep it for them."

Good luck was bid, and Jack rode over to where I was at work. He towered on Foggy, and I was thankful that he blocked out the hot sun for a moment.

"How's it comin'?"

"Just fine," I said, and realized I was very thirsty from the hard sound of my throat.

"Ye need water?" he asked, already reaching for one of the four canteens that hung from his saddle like Christmas ornaments.

"Nope," I said, and went on working. "I'm fine."

"Cut yerself," Jack observed.

I nodded.

"It's nothing."

"Ye ought t'clean that jest the same," Jack told me. "Else it'll swoll up and drop off."

I followed his advice, being fond of my thumb.

Jack loped off back for the herd with his precious canteens.

We worked on into noon, the sun directly overhead, inducing sweat like an eagle's eye on a field mouse. Monday had gathered up our tally and joined us in the work. He offered War Bag water, but still the man would take none. I refused as well, though my throat felt like an old boot in the sun. I found a pebble and tucked it into the corner of my mouth, as Roam had taught me.

Around five o'clock Fuke and Jack came up on their horses, looking weary.

War Bag called out in a strong voice, still working at the same pace:

"What's the Gospel, boys?"

"According to Fuke it's near *eighty,*" said Fuke, leaning across the horn of his saddle and cradling the still smoking rifle in his arms. "According to the *nigra,* sixty or so."

"Where is Roam?" War Bag asked.

Jack gestured behind him to the plain, which was spotted with dead buffalo.

"He spied another bunch off southwest. Them buffs is shore spread out," Jack observed, his teeth brown with chaw.

"Hey *Sin Buster,*" Fuke called to Monday. "What's for lunch?"

"Whatever you want to cook!" Monday retorted. He was loading up his wagon a ways off to the east of us.

Fuke groaned.

"Now will someone tell me what in the *hell* use he is? I'm famished!"

"Maybe you can go fry up that goddamn cat," War Bag said.

Fuke grinned and straightened up in his saddle.

"I'll save you some," he said, and kicked Napoleon full in the ribs, sending the Appaloosa bounding off over the prairie toward camp.

"Ye best leave that cat be, Fuke!" Fat Jack roared, and spurred Foggy after.

The old nag resented the gesture and bucked the big Missourian clean from the saddle, then made for camp on its own.

War Bag and I both laughed out loud till our sides were set to burst.

Jack sat up slowly and rubbed his tail bone with one big hand.

"Are you alright, Jack?" I asked, smiling.

"That sack of glue can lay tracks when he wants to, I guess," War Bag said.

"He ain't hardly used to the saddle yet, I reckin," Jack muttered.

"If I was a horse, I'd ruther pull a plow than carry you around, Fats," War Bag said.

Jack just shrugged and limped off after Foggy.

War Bag sat down on the carcass he was working and scratched his chin. His beard was flecked with blood and his clothes and hands were gore splashed. The fierce disapproval was gone from his face, as though the slaughter had cooled his anger. He nodded to me in a friendly way.

"What's your tally, Stretch?"

I took out my ledger and thumbed through it, leaving red fingerprints on every page.

"I got eleven so far," I said, trying not to sound proud. "What about you?"

"Nineteen," he said, fingering the scar above his eye.

He was a machine.

"Be a shame for you to cut out now that you're startin' to get it," he said slyly.

My bones ached. My skin was red, and my fingers grimy. But I did not want to leave the outfit anymore.

"Yeah," I said. "I guess."

I desperately wanted a drink of water, but Jack's fool horse had run off with the canteens, and I had neglected to fill mine that morning.

When I heard thunder I was ready to bless it. I felt the tension in my limbs subside, as though my whole body were sighing in expectation of the coming relief. It was a low, long rumble, and reverberated even through the very ground. I shaded my eyes. The sky was clear and fine.

Then War Bag yelled,

"Stampede!"

Third

The Buffs Came Running from the south, a vast dark carpet of heaving humps, maddened eyes and glinting horns that broke suddenly across the horizon and consumed the ground before it like Behemoth come to claim the Devil's own. They seemed to hover on a blurred cloud of shorn grasses and rolling dirt. Maybe they were in the all encompassing bovine panic that swallows the scream and the low and rolls the eyes back in the skull, or maybe their voices were lost to our ears in the cacophony of their thrumming hooves. Otherwise, they were quiet as death in the night.

The horses reared in their hobbles and stumbled, trying to get away. Othello got loose and bolted.

The buffalo spread from east to west, and cold fear ran down my body as I saw there was no escaping them. They gained ground with a rapidity that hypnotized me. I could see the froth streaming from the lead bulls' mouths in translucent whipcords, leaping and diving in mad patterns that caught the sun in quick little flashes.

War Bag was shouting, but his words were lost to me as the thunder of a thousand hooves grew louder and nearer. He grabbed my arm and jerked me down behind the carcass he had been working.

I fell on the half-skinned animal, smearing my face in blood. I saw Solomon's tether lying severed nearby.

War Bag raised Roam's Spencer rifle to his shoulder. Though the buffalo filled our vision, he stood still, aimed, and fired.

He shouted down at me, draining the carbine as he yelled and not taking his eyes from the buffalo. It was the only sound I could make out above the thunder that seemed to vibrate not only in my ears, but through my head and shoulders, down the ridge of my spine, and under the balls of my feet.

"Shoot!" was what he screamed. *"Shoot!"*

I wanted to tell him that I had left my pistol in Monday's wagon for safekeeping. It would have proved to be of little worth had it been in my hand then anyway, for I could not raise my head above the carcass. The sound, the force, it pinned me to the earth as sure as gravity.

But he was not shouting at me. Jack ran up beside him with the Henry rifle just as the Spencer went empty.

War Bag put up the barrel and fed the hungry carbine from his cartridge belt. Then he leveled it again and fired mechanically into the onrushing mass, swiveling the weapon left and right and raking the front line of approaching buffalo. I thought he was taking random aim (Jack surely seemed to be, firing in a panic as he was), for with so many animals what good would accuracy do?

Directly in front of us, through the cloud of gun smoke and rolling earth, one great bull flinched as War Bag's bullets struck him, exploding in little splashes of red across its fearsome face. It lurched and plowed headfirst into the ground, continuing for a few feet on its chin before coming to rest only yards in front of us. Almost immediately another fell right beside it, and then another to the left.

This happened precious seconds before the herd reached us. As the lead animals fell violently, two anxious followers stumbled over the bodies of their leaders, driven into the makeshift barrier by their fellows. Five or six buffalo collided and caromed into each other right before our eyes. They flipped head over heels and came crashing down bellowing on their companions. Some rolled aside and jumped up shaking their massive heads, only to be bowled over again by the ones rushing up behind.

As a result of this pile up, the herd parted and flowed around the trouble spot. The three of us crouched behind our skinned buffalo, a delta in a dark river of moving animal flesh.

They were an unending blur of brown fur and black horns, not more than six or eight feet on either side of us, snorting as they ran past. The sheer power of the stampede withered me, and I leaned on the dead buffalo for support. Jack genuflected beside me with his palms flat on the carcass, awestruck. His hat was gone, and the palm of his left hand glowed red, seared from gripping the barrel of the overheated Henry.

War Bag stood above us both, his hat hanging from his neck, his long hair escaping from its confining braid in wild, wiry bunches. He let the rifle fall. He pursed his thin lips, disapproving of the stampede. He seemed to glare at each buffalo as they shot past him, as though marking their faces for some later reckoning.

It finally passed nearly forty minutes later.

The ground around us was disheveled and beaten, riddled with hoof-craters and littered with trampled calves and the bodies of those buffalo too slow to keep up. Nothing stirred in that terrible wake, and the rumbling faded into the distance.

Jack and I slowly got to our feet and surveyed the waste.

"Ye reckin the others're awright?" Jack said.

I looked to where I'd last seen Monday's wagon. There was no sign of him.

War Bag walked over to one of the bulls he had shot and stooped down to examine him. He pulled something from its neck.

It was the feathered shaft of an arrow.

"One of Roam's?" I asked.

War Bag said,

"Nope."

On the walk back to camp we found Foggy. The Missourian had managed to recapture the old horse and walk him most of the way back to us when he'd realized the danger, taken the Henry rifle off the saddle, and made for us. Foggy had of course wheeled about and taken off again in the opposite direction.

But the buffalo had been too fast. We found the old horse where they had overtaken him, gored and beaten into the earth.

Jack knelt beside the gray and looked dejected.

"Ol' Foggy was my paw's hoss," he said sadly.

War Bag and I left him alone to gather his rig and what gear he could salvage.

War Bag turned the arrow he had found over and over in his fingers. Pete Adderly's evil warnings danced through my terrified mind, and I imagined an army of feathered savages coming to ride us down.

"Do you think the others are safe?" I asked dumbly, like a child needing empty reassurance.

He shrugged, and had a look on his face that said it didn't matter what he thought.

Jack rejoined us, bearing his saddle on one shoulder and three canteens on the other. We were covered in dust and the warm water was welcome. I spit out the pebble that had been clenched in my jaw for so many hours, and thanked the Lord I had not swallowed it in fear during the stampede. It would've been a fine joke to survive such harrow and then choke on a measly pebble.

The camp was in shambles. The hundred or more yards of tanning hides we had left there that morning were scattered all over the prairie, and in such a state as would not have warranted gathering. The Dutch oven was overturned in the unlit fire pit. Our meat vats had become pitfalls. Two buffalo still lay in the smoking vats, drowned head first in the spoiled brine.

War Bag found Solomon, and caught him without much difficulty. The old man checked the horse for wounds, and finding none, patted his flank fondly.

I saw nothing of Othello, or of Monday and the hide wagon, but the camp wagon and the team of mules stood a few feet out of camp, miraculously unmolested.

"I'll be," Jack muttered. "How d'ye s'pose they come out awright?"

The mules stood and looked at us, twitching their ears or shaking their brushy manes.

"Where's Fuke?" I asked, a little worried.

No one answered.

We heard the crack of a light rifle then, and Fuke came loping up from the west. To my delight, he was leading Othello behind.

"'Lo, boys!" he shouted, when he was close enough for us to hear him.

War Bag raised his hand in greeting.

Fuke rode into what was left of camp and tossed Othello's reins to me, smiling grandly.

"That was a *sight*, wasn't it?" he said, dismounting and grinning like the devil. "How about it, *Juniper?* Still think you could've outrun 'em?"

"How about Roam and the bull wagon?" War Bag called to him.

"They're back *there* about a mile. That muleskinner *tipped* the wagon, so they're trying to get them bulls *untangled*."

"Never let a muleskinner whack a bull train," War Bag muttered. "Why ain't you helpin' em?"

Fuke shrugged and smiled.

"I don't know *shit* about bulls."

"Let's all of us go and learn," War Bag said.

We spent the better part of an hour arighting the bull wagon and rehitching the train.

Monday told us in a jittery tone how he had overturned the bull wagon in his flight from the path of the stampede. When he saw his mules unhurt, he got down on his knees right there and said a loud prayer of thanks to the Divine Dispenser of Boundless Mercy. Roam had whipped poor Crawfish into a frenzy and high-tailed it up an embankment.

"Then all's I had to do was sit up there for a while an' watch all them big shaggies go by," he said.

"You had it *easy*," was all Fuke had to say about it.

War Bag showed him the arrow he'd found.

"Ain't s'posed to be no Indians runnin' buff 'round here," Roam said.

"Somebody ought to tell them."

"Are they renegades?" I asked.

"Sometime the Indian agents let the well-behaved Indians run buffalo," Roam said.

"Most times they don't," War Bag said.

"I seent another herd further south," Roam said.

"Might be better to go after that one than chase down these bastards all over," War Bag said.

"My way of thinking,'" Roam nodded.

"Well, let's gather up our kit," War Bag sighed. "We can head after them in the morning."

"*Swell*," Fuke said.

I had to admit I was perturbed at losing much of the tally we had worked so hard on, but War Bag assured us that he would compensate us for the work we had done out of his own share of what was yet to come.

The night's sleep was fitful. We didn't know if the Indians who'd stampeded the herd were still out there. Roam said they'd most likely run after the herd we'd been hunting, but I noted he sat up late into the night with his rifle just the same.

Fourth

We Found The Dead Men three days later.

We weren't more than a day or two on the trail of the herd before Roam made a discovery that disheartened us all.

"Wagon marks," he said, stooping in the trampled grass and pointing to a set of parallel lines that cut through the packed earth as plain as blood on snow. "Ain't but a few hours old."

War Bag tipped his hat back and scratched his forehead, sighing heavily.

"Well, let's go see what they got to say."

Our morale was dwindling. Already we bore the burden of having worked the previous herd to no avail. Three fourths of our tally had been lost to the stampede, and now another prospective herd looked to have been staked by a rival outfit. Fuke muttered harsh words to the effect of shooting the next Indian he saw. Monday was in despair, and feared his wife and child would starve in Kansas before he had enough money to take home. Even Jack seemed melancholy, and I heard him utter a bootless regret that he had traded his grandfather's Hawkens rifle for the big bent knife from the East Indies.

The wagons (there were two) had come from the direction of east Texas. Roam told us that they were unburdened, and were set in the wake of the buffalo. No hope of them being pilgrims or freighters. They were runners, intent on killing.

We moped along for three hours before we caught sight of a plume of smoke and crows.

"*Camp* fire," Fuke said with a trace of bitterness. "Maybe we'll be in time for lunch."

Roam peered up at the sky and said,

"Awful lot of crows."

While butchering and skinning our buff we had attracted a lot of crows too. Fuke would sometimes sit in camp and shoot them down for sport.

"Their fire ain't so high," Roam observed, "for all the smoke."

The prairie offered no sound but the chorus of insects.

The little camp took shape as we approached.

"Looks like they're having some trouble with their wagon," Monday said.

It was true. Of the two bulky shapes that were slowly becoming wagons, one was on its side.

War Bag, who was in the lead, stopped his horse and looked all around. He took out Roam's Spencer rifle. We all armed ourselves. The empty land only stared back at us, silent.

Roam peered ahead, leaning forward in his saddle as though bringing his eyes a few inches closer would help. Apparently it did.

"That ain't no camp fire," he said, his pistol in hand.

War Bag said,

"No it ain't. Let's both take a look. You boys stay here and keep your eyes open."

We sat tense in our saddles as War Bag and Roam cantered ahead.

My eyes hurt and watered with the strain of trying to see what they had seen. Only the pillar of smoke and the wheeling crows were clear to me. There was a familiar smell in my nose.

Jack, who was all the way in back with the wagon as we were traveling single file, shouted ahead.

"'Lo! What's the matter?"

"Put a cork in it awhile, *Fats!*" Fuke shouted back at him, turning Napoleon about nervously, as though he expected at any moment an attack from all sides. His Winchester was propped on his hip.

We all watched as War Bag and Roam reached the camp. Instead of entering right away, they circled it cautiously.

Then Roam came up alongside War Bag, dismounted, and walked in. No one rose to greet him.

"I don't see any animals," I proclaimed after a bit. "Do you see any animals?"

Other than War Bag and Roam, there was no hint of movement in the camp.

"*Quiet,*" Fuke hissed.

I looked at Monday. He shook his head no.

Roam fired the Spencer at the earth. Two dogs of some kind bounded out of the camp in a hurry, yelping all the way.

After a bit, War Bag raised his rifle over his head.

Fuke clicked his tongue and Napoleon started forward.

"Come on," he said to us.

Little details of the camp began to sharpen as we approached. I saw that not only was one of the wagons on its side, there was a team of mules lying dead in the traces.

Pots and pans were strewn about, as they had been in our camp after the stampede. The earth was battered with hoof prints, but of horses, not buffalo.

There was something else strange, flitting through the air. What I first took to be airborne dandelion seeds were feathers. Enough to fill a dozen goose down pillows.

Off to the side was a heap of turkeys. There must have been eleven or more, just thrown together. Most of the big birds' breasts were completely blown apart, as if they had been clipped by buffalo guns.

"Who'd use a big fifty to grease a buncha turkeys?" Jack muttered.

"Goddamned pumpkin rollers, likely," War Bag said. "Ain't a bit of sense among 'em."

Roam was picking through the wagon, and drew out an old blanket. He stepped over to the source of the smoke, which we had all presumed was the fire pit, and began beating it out.

I realized it was a body he was standing over. The thick smell was that of burning flesh.

I put the back of my hand to my mouth as I felt the bile rise. Sure in my mind that it was Indians, I forced myself to stay with the others. The prairie was suddenly too open. It gaped all around like a great mouth threatening to swallow me.

"Good Lord," Monday whispered.

Between my legs, Othello got jittery and shook his head, not wanting to proceed. I got down and led him by the nose. The familiar smell was that of death. That thick, unbearable smell every man identifies at least once.

"How many?" Fuke asked, when we were in speaking range.

"Two men," Roam said. "The mules are dead too."

Something like ice water broke at the nape of my neck and trickled down to the small of my back. Roam laid a blanket over the blackened bodies. They had been stretched out spread eagle on their backs. Stakes driven through their wrists and feet bound them to the burnt earth.

The mules had their throats cut, and their hams were cut away.

"Their guns and powder are *gone*," Fuke muttered, kicking over a smallish kip pile. "Unless these boys kilt their buffs with good *intentions.*"

"Here's another one," War Bag announced.

A few feet out of camp one of the hunters had tried to run and been pinned to the earth by arrows, some of which were stuck in his back up to the feathers. His scalp had been taken.

"Cheyenne," said War Bag, pointing to one of the arrows he took and brought back to show us.

He tossed it to Roam, who caught it, and squinted at the markings.

"Dog Soldiers," War Bag said, and I remembered their old argument at Cutter Sharps.' The old man looked at Roam, almost accusingly.

"And these're war arrows," Roam said, turning the missile in his hand before letting it fall.

Monday got down from his wagon and stood over the bodies, mumbling prayers from his Bible.

"Holy *Christ,* look at this one," said Fuke.

Fuke was standing near the broken framework of the hunters' hide drying rack. The wild dogs had pulled it down, and it lay in a tangle. In the midst of it all was the remains of a fourth man. He had been skinned, and his epidermis stretched out on the drying rack exactly like an animal's.

Or a buffalo's, I thought, hearing Pete Adderly's warning in my ears.

Though I jerked my head away, perhaps thinking to fling the bloody vision out of my mind, it has remained with me every night of my life.

"This didn't happen too long ago," Roam said.

"I'll get a spade from the wagon," Monday said.

"No," War Bag said.

"Lord! What have you got against burying people?" I shrieked, my voice cracking. The cold sweat of death-fear was about me.

War Bag sat on his great horse, grimly looking at all of us.

"You boys all heard Roam. Whatever sons a' bitches 'did this are probably nearby. We're going to break for Adobe Walls," he said, jamming his finger southeast. "If we keep at it, I'm *sure* we can make it inside of three days. Now, there's nothing you can do for these men. They've already been picked over and no amount of dirt is gonna change that now. We're headin' out, and if you wanna waste time diggin' holes for 'em, you may as well dig some for yourselves too. "

He looked at each of us, and as his eyes passed over me, I felt tiny needles prick my arms.

"Now saddle up," he said.

That ride was a horror. Every shrub we came across seemed to hold Indians. Every dip in the land hid an Indian pulling back a bowstring. We saw two herds of buffalo that day, but passed them by.

We rode until the night was black and the only way to see Roam and War Bag ahead of us was to lay flat and skylight them against the stars.

When at last we stopped to bed down, we set a watch. We sustained ourselves on jerky and water, not daring to cook anything. We fed our animals blindly in the dark, and grew to know them by touch and sound.

Roam instructed us all to leave our mounts un-hobbled, and to sleep with one arm through the reins. If trouble were to come, Jack and Monday were issued standing orders to leave their wagons and double up. Monday with me, Jack with Fuke. Jack tied a string around the neck of his cat and fixed the other to his wrist. The accursed animal complained throughout the night until Fuke, who was on first watch, told Jack to let it go or he would smother it.

"Ye touch this heyar cat, Fuke, 'an I'll open you up," was Jack's monotone reply from the darkness. "You kin b'lieve that."

Fuke did not say another word.

We all had our hackles up, and the yodeling of the cat did not help things. Jack hushed it and tried to keep it calm, but Whisper seemed to sense our unease and was too stupid not to express his own.

After a bit War Bag said, quietly but quite firmly,

"Let him go, Jack."

Everyone was silent, though there was a change in the pitch of the cat's voice for a moment. I saw Whisper's sleek form dart swiftly from the darkness and then disappear again.

We did not hear him again that night, and no one spoke.

Roam woke me for the second watch.

We passed the hours in total darkness. Maybe because I was within the shadows I did not fear them. I knew my companions were all around me and that Roam, one of the most able of them all, was awake and watching the dark where I sat.

"Roam?" I whispered, not really because I had a question in mind, so much as I wanted to assure myself that he had not been dragged off under the knife of some silent Indian.

"Yup?" he answered, very quietly.

I voiced the first thing that came to mind.

"Where did War Bag get that scar above his eye?"

"Beecher's Island," Roam said.

"He was there too?" I asked.

"He was," Roam said.

"Was that where you met him?"

"Yup."

"He was a soldier, then?"

"He was a civilian scout when I met him—one of Major Forsyth's volunteers."

I leaned forward.

"So he got it in a fight with Indians?" I asked. War Bag's scar was an intriguing book cover illustration, and I wanted to read the story or have it told.

I heard Roam settle before he began.

"Beecher's Island was jest a little stand of mud and brush in the middle of the Arickaree Fork. I 'member there was one lone cottonwood growin' on the north end. You can't find it on the maps 'cause it ain't there no more. The river done swallered it up.

"This was roundabout six year back. The troop was camped near Cheyenne Wells on Sandy Creek, when two men ridin' dispatch outta Fort Wallace for Cap'n Carptener found a pair of Forsyth's scouts out on the prairie and in a bad way. Major Forsyth had mustered up a unit of fifty scouts to go out an' hunt Cheyennes sometime early September. That fall ever'body was huntin' Dog Soldiers.

"Well, them two white men tol' them troopers that Forsyth had got into trouble with the Cheyenne at the

Arickaree—that they'd been makin' a stand on that little island for goin' on three days. That was five days before them two had slipped out to get help. Once Cap'n Carpenter got word, he divvied up thirty of us and requisitioned a supply wagon. We made for the Arickaree at a gallop with the rest of the troop followin' behind.

"As I recall it was me an' old Rube Waller right behind the Cap'n, so we was the first to see them Injuns. They was near seven hunnered of 'em up on the ridge shootin' down at that island when we got there. The Tenth opened up on 'em, and they rabbited. We chased 'em for 'bout three miles before we called it quits and went back to see about Forsyth and them men.

"They was a sight. Eight days they'd been out there, diggin' in the sand tryin' t'get at water. They'd made a breastworks outta they horses, an' fell to eatin' 'em when they run outta food. 'Place stunk like the nation. They was flies ever'where. The boys all set to givin' 'em water and whatever rations they had. 'Course it made 'em sick straight off. Most of the men didn't know no better, and them scouts was starvin,' and jest couldn't help they selves. We had to get a hold of some of 'em, oncet the Captain gave the order to hold off on the water and food, as a couple of 'em threatened to kill us if we didn't feed 'em right off.

"But not ol' War Bag. I found him layin' between a dead man and a dead hoss. He was half-scalped, and when I went over to him, he jest asked me for my canteen to wash the blood outta his eyes. I thought it was a trick, so I done it for him, but he never once tried to take the water from me. He jest laid there real still and let me wash the wound.

"He couldn't talk too good, but after a couple little sips of water, he told me on one of the Indians' charges through the men, a Dog Soldier jumped him and tried to scalp him. He shot that buck in the face to get him off."

"Killed him?" I asked.

"I didn't see no Indian, but they carry off they dead."

I thought of the powder burned Indian War Bag had mentioned.

"So he might still be out there? Powder burned, like War Bag said?"

"War Bag crazy," Roam said dismissively. "Man live through somethin' like Beecher's Island, he get to thinkin' jest 'bout anything."

I imagined War Bag's nemesis. A black faced Indian itching to complete the work he started on our leader years ago.

"After the fight, we buried them that had died. War Bag

an' me got to talkin' 'bout this, that, and the other thing. We took 'em all back to Wallace a couple days later, and then we went on a drunk over at Pond Creek after they was up an' around. Them white scouts shore treated us soldiers. Lord, it was glorious." And by his tone, it was.

I dwelt for a long time on the sights and smells of the past day, and on the powder burned Indian. Jack and War Bag woke to relieve us and I fell asleep listening to the big Missourian call for his cat lowly in the dark.

No Indians, powder burned or otherwise, came to raise our hair in the night. Jack found Whisper lazing in the shadow of the hide wagon that morning, cleaning the blood of another field mouse from his whiskers.

We rode on. No voice was raised in song, and Monday read no Psalms to his mules. Like true angels, they were heedless of the danger anyway.

At one point we saw more smoke far to the east. War Bag said there was no point in us diverting our course to investigate. I was glad, for I was sure if we did we might stumble upon more fresh corpses, or worse. Fighting Indians had always seemed a romantic endeavor to me, but after the sights of the previous day and Roam's relation of the fight at Beecher's Island, I was quite sure that I was not prepared for an encounter with real hostiles. Perhaps Indians had no time for prisoners, but they surely made time for torture.

Around early evening we saw a large band of riders headed northeast. They were too far away for any of us to clearly perceive, but War Bag was sure they were Indians.

Camp was made with more trepidation than the night before. Even Whisper seemed to sense it, for he made no sound and did not stray from Fat Jack's side. We ate jerky and drank from our canteens again, and the fond times of hot coffee and fresh meat seemed far away.

I pined for a hot bath, a shave, and a set of clean clothes. The practice of dunking myself in river water was getting played out. At night I had trouble sleeping due to the feeling of tiny legs scuttling across my skin, both real and imagined. My companions were getting along in the unwashed area as well. I could hardly believe that our collective scent was not enough to repulse any human beings, let alone Indians, who I had been led to believe possessed a keener sense of smell than white men.

Perhaps I was right in this assumption, for no Indians came that night either. Just a bad dream about fire, and raving savages with ink smudges where their faces should have been.

Fifth

Adobe Walls loomed like the promise of a warm, waiting bed in the backs of our minds. We broke camp at first light.

All that day we spied signs among the grass of the passage of many horses and wagons. We kept on our guard, attaining a higher awareness born of fear. I could at times look straight ahead at Fuke's back and feel the country open up on either side of me, then slowly split to include Jack and the wagon behind, and some animal darting just out of sight to my left. The world was wide and Othello was turning it like a wheel, walking in place while I sat atop his broad shoulders and observed all that passed underneath, like Argo.

At noon, when the sun had ascended to the peak of the firmament, we came across a passel of dead horses. There were perhaps fifty or more. We were not as surprised by this development as we might have been, for the sky was filled once more with crows, and we had seen them a long way off. They were having quite a feast.

"Cheyenne ponies," Roam said, looking at the cracking paint on their swollen hides. "Some Comanche, too."

"What's thet thar red hand?" Jack asked. There was a ghostly red hand print on the right shoulder of most of the ponies, as though the devil had slapped his infernal blessing on every one as they rode off to murder.

"Means it's a war horse," Roam said. "Them ex-marks on the rump is dead men."

"They've been shot," Monday said, gesturing to a large bore bullet hole in the breast of one of the horses.

Fuke whistled lowly, and suddenly a shot rang out like a discordant note, kicking up the dirt very near to where War Bag sat on Solomon. The big black horse reared a bit, but the old man wrestled him easy.

We all went for our guns, but War Bag raised his Henry rifle and fired three times into the sky.

About two miles to the south I spied a spread of four or five small buildings and a corral sitting on the west bank of a little creek. That was where the shot had originated. As we watched, there came three more in answer, spaced evenly and fired into the air, the same as War Bag's.

One by one, men began to come out of the cabins. I was relieved to see the wide brimmed hats that meant whites.

As we came into Adobe Walls, we saw evidence of a great fight. The outer walls of one crude picket store were peppered with bullet holes, and just outside of the corral were a series of freshly turned mounds that looked like graves. A broken Studebaker lay on its side near one of the southernmost structures.

Thirty or forty men were about, all buffalo runners. There were lookouts on top of the buildings, crouched in little makeshift sod breastworks, rifle barrels slender and challenging, little heads darting back and forth. The windows of the buildings were barricaded with boards and flour sacks, and the sill of one picket store was dusted white where one of the sacks had been burst by bullets. I saw Indian arrows sprouting everywhere like wild daisies. Some shot up from the ground, others stuck in the walls, and some grew from livestock lying dead and swollen.

In between two buildings I saw stacks of buffalo hides in a little yard behind one of the stores, and part of a dead naked body sprawled behind a kip pile. I shivered.

"Who fired that shot?" War Bag bellowed as we dismounted in front of a picket house where most of the men seemed to be gathered. A bullet-riddled sign over the door of the place read, "Jim Hanrahan's Saloon."

A long haired, good looking youngster with a Sharps .44 and the dirty beginnings of a mustache raised a hand to us. He looked the coolest, and kept his head while the other men clucked at each other and gave a lot of orders no one seemed inclined to follow. There was a black feathered Indian lance leaning up against the doorway where he stood.

"It was just a warning shot. Didn't meant to scare you none. I'm Billy Dixon. Who's outfit are y'all?"

The others hushed as the youth spoke. He wasn't much older than I was.

"My outfit. I'm Ephron Tyler," War Bag said, putting out his hand.

Billy Dixon took it, visibly relieved that the old man was not offended.

"Well, I hope your boys can shoot, Mr. Tyler," said another man, a mustachioed Irishman with a red face and a big cap and ball pistol between belly and belt.

"What happened here?" War Bag asked.

Billy Dixon looked confused.

"Why . . . Indians! Four days ago they come out of the hills to the east. Hundreds of 'em. We fought 'em off, but they're still up there. Didn't you see 'em?"

Hundreds, I thought, looking all around me.

"We didn't see no Indians."

"You're lucky to be alive yet," Billy Dixon told us.

"Luckier than *some,* that's for certain," Fuke mused.

When the men looked to War Bag for explanation, he said: "About two day's ride to the north we found a camp of white men dead and burned," War Bag told them. "We believe it was Cheyenne Dog Soldiers."

"Not just Cheyenne. Comanche, Kiowa, and Arapaho too," the Irishman said.

"How many men you lose?" Fuke asked.

"Three dead white men to thirteen or so Indians," said Dixon.

"Are the men buried?" Monday asked.

"All but the Indians and a nigger bugler that rode with 'em," Dixon answered.

"*Bugler?*" said Roam.

Dixon slid his eyes briefly to Roam, sized him up, and then gave his attention back to War Bag without saying a word.

Roam looked at War Bag.

War Bag spoke up,

"What's this about a bugler?"

Dixon turned to one of the other men, a weasely looking sport with a carefully styled foot long mustache sitting on a cut log outside the saloon.

"Go on, show 'em, Harry," Dixon said.

Harry looked around at us nervously and produced a shiny brass cavalryman's bugle from the ground beside him. He stood up and displayed it to War Bag, turning it over in his hands, but not letting go, as though it were a treasure. It was well cared for, and there was a rawhide loop strung with trade beads passed through the stem. There was an etching near the mouth that plainly read: *10th United States Cavalry.*

"Where'd you pick that up?" War Bag asked Harry.

"Off'a nigger I kilt," Harry said defensively. He gestured to the overturned Studebaker near a smallish picket building that looked to be the stockade. "Got him over there, whilst he come a'runnin' out with a sack of sugar and a coffee can."

"He was sounding charges and rallies all that first mornin,'" said the Irishman. "I was in the cavalry, so it helped us a bit, knowing what they were fixing to do by the bugle."

Roam had been staring intently at the bugle since its appearance, and now he nodded his head and excused himself. I saw him head for the wagon where the Negro bugler had fallen.

Curious, I tied my horse to Monday's wagon and went along after him. Monday jumped down from the driver's seat and went with me, the voices of the men growing smaller as we caught up with Roam.

The buildings were staggered in a north-south line, with the corral and stockade on the southernmost edge. Had the bugler fallen closer to the stores, they might have buried him for sake of the stench. He'd had the bad luck to die away from where the men slept, and so had been left to rot in the sun. I was not sure I wanted to see the corpse, nor was I sure why Roam did, but I went along anyway.

We saw him at last around the corner of the overturned wagon. He was sprawled face down, his hips pivoted and his arms twisted about him, the pilfered sugar sack still under his arm, and split open. The white sugar was scattered with the dirt and stained with . . . blood? No. Not unless blood moved. It was crawling with ants.

I halted, and Monday took two steps and stopped, turning to me.

"I don't believe I want to see him," I whispered to Monday. My mouth had gone dry.

Monday nodded and continued on till he stood beside Roam, who was already looking down on the corpse.

Roam lifted his bandanna to his face and sat down on his heels.

Monday took out his bible and read quietly.

I could hear the men talking in the distance, and the quiet murmur of Monday's offerings for the dead man's soul. I looked over at the corpse, which was just far enough away to remain discreet.

The dead Negro wore beige wool breeches tucked into black leather cavalry boots. It was the only sign that he had ever lived among civilization. Above the waist, all the rest was savage. He was barechested, and had a necklace of rawhide and bone hung around his neck, along with a breastplate of porcupine quills that clattered when Roam reached down and turned him over. He was stiff and dry as a log.

Roam straightened up and took off his hat. He swatted at the body, but not disrespectfully. I can only assume he was shooing away the ants that had set upon the dark flesh.
Monday continued to pray. Roam walked over to the wagon, looking inside.

"Who is he? Do you know him?" I called.

Roam reached into the wagon and took out a blanket that had been left in the box.

"I knowed him oncet," Roam said, spreading the blanket on the ground beside the body. "Arlo Flood was his name. We was in the Army together."

Monday closed his bible and helped Roam move the stiff and bloated body onto the blanket and swaddle it.

I felt foolish for standing by like a prim gentlewoman. I came over, and that's when one of the buffalo runners from the saloon walked past me and came to stand over Monday and Roam.

"What do you think you are doing?'" said the man. He was a short Frenchman, with skinner's knives shoved through his belt. He looked flustered and contrary.

"We're burying this man," Monday answered.

"The hell you say! Let the buzzards and the wolves have that son of a bitch."

"He deserves to be buried," Monday said firmly, without a tremble.

"He deserves to be shat out by a dog," said the skinner, spitting into the dust in front of Roam.

Monday just looked at the Frenchman. Roam turned to face him.

The skinner put his hand on a flap holster that encased a pistol on his hip. I stood behind him, ignored.

"You leave heem there," the Frenchman said.

I touched my Volcanic, feeling my fingers curl around the butt. I was trembling all over suddenly, and felt a dam of some kind break at the base of my scalp.

"You go to hell!" Roam yelled at him.

The small man took out his pistol.

My own hand fell away from the butt of the Volcanic and I stood transfixed, my shoulders tightening.

Then a shrill cry came from back by the saloon.

"Indians! It's Indians!"

All of us stood frozen for a half an instant.

I saw the men running pell mell for the store and the saloon. Jack and War Bag paused in their flight to look over at us. Fuke disappeared inside.

I looked up the rise to the north from whence we had come. There was a mass of bright shapes bouncing on the ridge like a multi-colored caterpillar slinking among the low hills.

Sixth

Then We Were Running. I was beside the Frenchman, our enmities forgotten, Roam and Monday sprinting behind.

We reached Hanrahan's, and got in before someone slammed the door shut. The men were all pushing their rifle barrels through cracks in the picket walls. Thin slivers of dusty sunlight intersected everywhere, like plotted geometric points.

I heard Jack call for his cat, and caught Fuke's voice yelling about the animals we had left outside. War Bag's deep bellow was unintelligible but authoritative. Everyone was talking at once.

Roam slid his carbine through a space in the wall. War Bag came up beside him, and the young Billy Dixon knelt there too, saying something about the store.

There was a ladder in the center of the room that led up through a fresh dug hole in the sod roof. The Frenchman scurried up it and disappeared.

"Wait! Wait!" War Bag was saying.'

"Where are the Indians?" I asked.

"I don't think they're going to attack," said Monday.

"Don't bet on it, son," some eavesdropper said.

Then I heard the singing. It was far off, and seemed to rise and fall like the gallop of a horse. It wafted through the air all around us, and plunged the room into an unnerving silence. It sounded like the wailing of ghostly animals. Like birds killed in flight and poisoned wolves, and something else. Buffalo. It sounded like the ghosts of murdered, flayed buffalo. Not the bellowing animal noises they made, you must understand, but the sound I believe their souls must make. Buffalo souls singing.

It rose in pitch, and you could follow the position of the Indians by the sound. I could not even hear Monday's breath and he was beside me. All the men looked as though the walls would collapse, as Jericho did under Joshua's trumpets.

Then it slowly dwindled, and was gone.

"They have crossed thee creek," the Frenchman shouted down from the hole in the roof.

"Gone?" said Billy Dixon.

"No," War Bag said. "Not yet."

He settled back then, lowering his rifle.

"But they're leaving."

"God help any man out on the range this day," said the Irishman.

War Bag said,

"Any man not Indian."

When the danger had passed, Roam and I went off alone and buried Arlo Flood.

The digging was slow work. I felt like a traitor with so many who had lost their friends in the attack standing in the doorways and windows nearby, watching and muttering. The French skinner especially seemed to resent our proceedings. I felt sure he would dig the body up when we had gone, and leave the corpse for the dogs as he had wanted. I do not know what kept the men from interfering. I suspect it was War Bag.

When it was finished, we had a dry hole four feet deep and six feet long. We rolled the blanket-wrapped body of Arlo Flood, Roam's onetime comrade, into it and quickly covered him. The sun sympathized with the runners, and punished us for our work.

I don't know why I helped Roam bury the renegade Negro. I don't know that he was a good man. But I was fairly sure that not even a bad man deserved to be shat out by dogs, or left in the sun for carrion.

The day aged slowly, the sun rolling up to the summit and then tumbling back down like the stone of poor Sysiphus in Hades. The men were tense, all eyes expecting the Indians to come charging in at any moment.

War Bag assured us that they wouldn't come at night, if at all, and Dixon agreed. We all retired to Hanrahan's.

Each of us drank deep at the Irishman's encouragement, for he assured us he was not prepared to lug whiskey barrels in his inevitable flight to Dodge City. Our money apparently posed no burden, for he charged us not a cent less than was posted. Billy Keeler, an old cook from Myer's store next door, brewed coffee.

It seemed it was everyone's intent to head north.

Billy Dixon swore he'd never hunt buffalo again, and a young man named Billy Ogg said he would buy a train ticket east as soon as he arrived back in civilization as he knew it.

"Damn shaggies ain't never worth half the time and trouble it takes to find 'em and kill 'em," Ogg said.

"'Specially not this time," said a strapping young skinner who spent most of his time staring out the doorway in the direction of the graves.

The old cook came up beside the young man and offered him a steaming cup of coffee. I smiled to myself. Billy Dixon, Billy Keeler, Billy Ogg . . . . Fuke remarked to me that there were more 'Billys' in this outfit than there were grey hairs on Jack's cat's ass.

Most of our band was silent, dreading the prospect of returning home penniless from Texas, where we'd been told that the buffalo flowed like milk and honey. I am sure the wasted time and money weighed heaviest on Monday. Fall would come, and then winter, and he still had little money to take to his family in Haskell County.

I wondered about my own future. Where would I go? The prospect of returning home seemed intolerable, but I had no money. I had gone into this venture empty headed and would return empty handed, it seemed, if I returned at all.

Jack opted for the duty of lookout, and went up on the roof. Soon the bouncing sound of his Jew's harp was floating down to us as he plucked out 'Arkansas Traveler' and watched the sun sink. Whisper lay in a corner of the room, sprawled on his back, asleep. His hind stump twitched as though it dreamed of the lost limb.

The rough little French skinner (we knew him only as 'Frenchy') seemed unconcerned with recent happenings. He sharpened a big knife on a strop and said little.

"I think I'll go back east," said Harry Armitage, the man with the mustache who had killed the Negro bugler.

"Where are you from?" I asked.

"Warrenton, Virgina," he said. "How about you?"

"Chicago," I answered.

"Really?" he asked, pulling his stool up a little closer to where I sat beside Fuke. "Tell me something about it. I've always thought to visit there one day."

"What do you want to know?"

"Were you there for the fire?" he asked. "The big one they had a couple years back?"

I shrugged. I wanted to accommodate him with a story, but my family had been in Iowa City visiting cousins at the time, and had only seen the aftermath.

"No," I said.

"Humph," he muttered. "That's a pity."

"What's for *us*, War Bag?" Fuke asked the question the rest of us had been biting back.

Roam and I looked to him. He sat in a corner, smoking his pipe, his rifle at his side and one foot propped up on a chair.

"You boys can decide what you want," he said. "But as for me, I'll be hunting buffalo."

"Well then you're crazy, mister," said Keeler the cook, putting down his coffee pitcher. "Tell me your name and I'll carve it in the bar. It'll be the only marker you're liable to get out here."

"It's Ephron Tyler," War Bag told him. "You get your knife whet, sheffy. 'Be a lot more thinking like I am."

But the old cook had stopped dead at War Bag's words.

"What'd you say your name was?" Keeler asked, coming over.

"You deef?" War Bag rumbled, looking at the grizzled cook. "I said Tyler. Ephron Tyler."

Keeler put his hands on his narrow hips and his voice dropped.

"Well. Then you had a son named Billy."

I had not quite had the chance to register what the old man had said before War Bag answered,

"I have a boy . . . ."

"*Had*," Keeler reiterated, tossing a thorn of a thumb violently toward the front door and the young night outside. "He's buried out there by the corral with the rest. And you can go to hell, Eph Tyler."

He spat on the floor right in front of War Bag.

Seventh

Silence Was All The Reply War Bag made. I expected him to raise up his terrible voice, to grab his rifle and let it explain. Any calamity or sudden violence could have ensued and I would have welcomed it.

But War Bag just sat there, thunderstruck. His skin hung low in the lamplight, and his eyes seemed to sink under our expectant gaze. His shoulders sagged, supported only by the back of the chair. Where was the defiant veteran Indian fighter and two fisted sourdough who moments before had declared his intent to keep to his course regardless of what any two hundred bloodthirsty braves had to say about it?

War Bag's boot came off the chair where it had rested and thudded heavily to the floor. He looked as though he had been shot with old Bullthrower itself. His chest was sunken, his face hangdog, his lips gone dry.

I put my eyes on Roam for want of something, anything else

to look at. I felt as though I were gawking like some rubbernecked passerby witnessing a private dispute on the street.

War Bag reached one enfeebled arm up and his long fingers touched the crown of his battered hat. He took it off and looked down into it, as though some explanation might lie within. His hair was in disarray, and the scar that wound through his scalp glared at us, as ugly as the moment itself.

All the while Keeler stood there with his hands on his hips, eyes glowering in the way only those that belong to bitter old men can. His apron was egg-stained and ridiculously inappropriate.

The young skinner to whom Keeler had given the coffee turned from the window and spoke quietly.

"Mr. Tyler, I knew your Billy. He was my friend. My name's Masterson."

War Bag's hands searched his hatband.

Roam looked at me, and I looked away. Not even Fuke had anything to say.

"He was with my outfit," Masterson said. "A swell shot." He took off his hat. "When the Indians attacked, he and Fred ran to the stockade to see about the horses."

Fred Leonard, an Englishman with protruding teeth, spoke up,

"That's the God's honest truth of it," was all he had to say, before scratching his neck uncomfortably and saying nothing more.

"He stopped for a shot in the doorway when he . . . caught it," Masterson said, swallowing.

Then War Bag asked, in a throaty voice,

"Was it quick?"

He did not raise his head, instead directing his question at the hat in his hands.

Masterson lowered his eyes trying to find an answer on the floor.

Merciless Billy Keeler saved him the trouble.

"No it wasn't. He got it through the lungs."

"Let the man be, Keeler," Jim Hanrahan begged.

"Hell," the old cook said, still glaring down at War Bag. "You had yourself a fine boy, Mr. Tyler. And he caught himself a bad end out here looking for his paw. Lookin' at you, it don't seem to me that you're worth one runny shit let alone a game young man like that one lyin' out there."

War Bag stood, and I think the walls and the floor and everything and everyone in between held its breath.

But he just put his hat back on his head and stomped out into the night without a word.

Keeler watched him.

"Go on, take a look!" He shouted after. Then, more quietly, "Take a good, long look."

"Talk like that is liable to get you shot, old man," Jim Hanrahan warned.

Keeler waved the Irishman off and made for the coffeepot.

"You boys best clear out of that man's company," the old cook warned us as he poured himself a cup. "Nothing good'll come from stayin' with him."

Monday and I were dead silent, reeling in our seats. I felt like I was in the middle of the ocean and I had just watched my compass slip over the side.

"Hey old man, get me some *coffee*," Fuke snapped at the cook.

"Get it yourself," Keeler replied, and walked away.

"What kind of *hash slinger* have you *got* here?" Fuke said to Hanrahan, as he got out of his seat.

"Och," he replied, with Hibernian brevity.

Billy Ogg shook his head and said,

"Old Keeler's just put out on account of he lost his dog."

"Yah! Tventy bullets in dat dog," said Andy The Swede.

"Only one for Billy," Masterson said sadly into his cup.

I turned to Roam.

"Did you know?"

Roam shrugged.

"Ain't somethin' a man's likely to talk about."

"He went all over the goddamn prairie looking for that man," Masterson said bitterly, looking out through the pickets. "Told me he hadn't clapped eyes on his paw since he was three years old. He said his maw had to run a beet farm in Colorado all alone."

I went to the door.

War Bag was standing over the graves. As I watched, he put one hand on the post of the corral and leaned on it slightly.

"Sounds like ol' Billy should've stayed *home*." Fuke remarked, pouring his own coffee.

Monday stood up, indignant.

"You mean to say you think it was alright for him to abandon them like that?"

Fuke sipped, considering his words.

"All I'm saying is that once a man *quits* his family, no son he's never taken the time to know is going to *convince* him to come back."

"I don't think Billy wanted him to come home," Masterson said. "I think maybe he was just looking for him . . . to know who he was." Masterson shrugged and pursed his lips. "Or to let him know who *he* was."

"Well, it sure didn't do him any good," Fuke said.

Monday came up beside me.

"It's horrible," he whispered, and for a moment I was not sure if he meant for me to hear. "A horrible, *horrible* thing."

"Where's *your* wife, sodbuster?" Fuke said.

Monday glared at Fuke, but said nothing.

Fuke smiled into his coffee.

I put my hand on Monday's arm.

He looked at me, his eyes sharpened by the accusation.

"I *am* going back," he told me. "Just as soon as I can get up enough money."

I just nodded, thinking of War Bag alone out in the night, over the grave. I wasn't sure how to feel about it. I felt a shred of jealousy at news of his having a son.

I think all of us in the outfit held War Bag in some sort of fatherly regard and struggled, however unconsciously, to be his favorite son. The news that War Bag had a true son, and a stranger nonetheless, had thrown us all somewhat. I could see it in the eyes of the others. Something had been lost. As for Jack, I was not sure if he had heard from his lookout's post on the roof.

The talk of the runners turned to exodus. Hanrahan announced his intent to head north for Dodge City, and several men opted to go with him. Frenchy, the small, mean skinner, said that he was not finished with Texas yet.

Our outfit was undecided.

"We're not making any money," Monday argued. "If we stay much longer out here we'll wind up scalped."

"I think with all these boys headed *out* of the panhandle, it'll cut the competition down *low* enough for us to still turn in a decent *haul* by winter," Fuke said.

I mumbled my agreement. It made sense.

"Gonna be a lot of hell with the hide off with them Indians still out there," Roam observed.

"You *heard* the old man," Fuke reiterated, rolling his eyes. "We won't be the *only* ones thinking this way. If we get into trouble, there'll be *other* outfits we can kick in with. If the old man's *willing,* I'm still game."

"You think he still is?" Roam asked, his eyes meaningful.

"Aw, that ain't got *nothin'* to do with us," Fuke said impatiently. "We signed on for *cash money,* and he knows he can't just cut us loose without *trouble.*"

Roam looked warily at Fuke, and Monday and I exchanged glances.

"Whatchoo mean by that, Louisiana?"

Fuke shrugged, embarrassed.

"All I'm saying is he wouldn't cut us *loose*. He gave his *word* to see us through the season, and that's good *enough* for me."

"There's somethin' else, though," Roam said. "We don't know the old man'll be in the right mind. He might take us out there with the idea of killin' Indians to settle it with his boy."

"How would killing Indians settle anything?" Monday exclaimed.

"Well, guilt be a powerful thing. 'Specially guilt like this kind here."

Fuke threw up his hands.

"Look, I'll face off red Indians, two gun *chili-eaters* and a whole horde of Philistines with the jawbone of an *ass* if it means killing buffalo too," Fuke declared. "I'll be *goddamned* if I go back to Dodge City with just my *fists* in my pockets."

Jack came down the ladder. He scooped his purring cat off the floor and came over. We told him of what had occurred.

"What you got to say, big man?" Roam asked him as he stood over us, his big shadow casting further gloom upon our proceedings.

"Well boys," Jack muttered, after a few moments' deliberation, "'reckin I'll go wheresoever the ol'man decides."

Fuke slapped his hand on the table, making the coffee jump in the cups. Whisper hissed and leapt to the floor, scratching Jack's thumb.

"Well dammit, I'm a runner and I'll *stay* a runner. War Bag or *no* War Bag."

"Awright then," Roam said. "If we get into trouble, I hope you got that jawbone handy, though," he said to Fuke.

"*Shit*, I'll just grab a hold of *your's* if we get into it," Fuke said.

We laughed, except for Fat Jack, who sucked his thumb.

Now all that remained was War Bag himself.

Jim Hanrahan let us sleep in his saloon (for a minor fee), and we spread out on the floor, taking turns watching for Indians on the roof.

During my shift the stars were out in formation on the heavenly parade ground. The land was empty and dark, and nothing human spoke through the long hours. I heard the creek rushing past the strip of cottonwoods, and as far as I know War Bag did not leave his son's grave all night.

He sat against the corral post, and had no more words for the dead than he had for the living. He did not cry for forgiveness or redemption. He just sat there, feet touching the mound of earth where a son he had barely known lay with strangers. Sometimes my sight passed from him, and only the occasional stir of his boots in the dirt reminded me he was there.

I pitied him as I had pitied no other man, and I feared him, too, and resented him all at once. He did not sleep during my watch. What does a man see who stares at a grave for so long?

All the men below made their preparations to leave in the morning, and I wondered if we would be going with them.

Eighth

War Bag Stood in the doorway of Hanrahan's with his hat on his head and his face shadowed by the morning sun. We had all slept late, full of Keeler's coffee and the Irishman's beer. He did not give us the chance to voice our concerns and opinions, nor tell him of the decision we had made to continue to follow him.

He only said,

"Time to go."

Then he went back outside.

Word of the Indians had spread. As Hanrahan, Dixon, and the majority of the other veterans were busy saddling their horses and hitching their wagons to leave, hundreds of other runners were filing into Adobe Walls for shelter from the plains that were said to be teeming with murderous Indians.

When we went outside onto the porch, they were coming from every direction; teams of oxen and mule freights six carts long, men with guns all talking excitedly, asking who had been here for the big fight and how many Indians were out and about. The Indians had brought death and fame to Adobe Walls.

We walked past most of the newcomers, letting the ones who were remaining behind field the questions, as most of them seemed happy to do.

When we walked out to our horses, Frenchy was staring at an Indian's severed head which he or someone else had fixed onto one of the corral posts. It buzzed with a halo of flies, and looked misshapen and unreal, like a burlap sack with a painted on face. When Frenchy caught sight of my disgusted grimace, he waved the palm of his hand over his mouth, pantomiming a war yell. Then he leered at me, and his teeth were sharp and white. It was a weird, carnivorous smile.

Soon we had the ponies saddled and fed. Monday was driving the mules south with Jack and the bull wagon in tow and the rest of us behind.

The Frenchman trotted up on a sorrel alongside War Bag. He spoke loud enough for us all to hear.

"I would like to come with you, Mister Tyler."

It came to us all as a great surprise, but War Bag nodded his assent.

"Twenty five cents a hide is what I'm payin.'"

"I will take it," the Frenchman said.

I saw Roam bristle as the little skinner fell back in line with the rest of us. He made no move at formal introduction, just joined us like an impetuous stray.

Billy Dixon and the others wished us luck as they began their long trek for Dodge City in the opposite direction. I saw Old Man Keeler standing in the doorway of Myers' store with some others, watching the comings and goings with a stone face.

War Bag did not look back, nor say another word to anyone for the rest of the morning.

The presence of the Frenchman weighed hard on us. He was a cruel, quiet little man, with coal eyes and sharp, angular features like a ferret's. His clothes were filthy, and his ratty slouch hat bore a drooping pheasant's feather that wagged like a wolf's tail as he rode. He spoke to no one that first day, but tied his reins to the back of the bull train and dozed in his saddle.

War Bag's welcoming of the Frenchman did not sit well. It meant more competition on the killing field. In addition, the man was a stranger and known to be belligerent toward Roam. Perhaps the old man was unaware of the near shooting over the burial of Arlo Flood, but to let a new man on without consulting the rest of the outfit ruffled all our feathers.

We veered southeast into the heart of Texas. Roam did not scout as far ahead as before.

The sun shined bright and the grasses swayed gaily in the first good summer breeze we felt that July. The land was in bloom and beautiful, and there was no room in nature for murderers red or white, nor for wayward fathers and their dead, questing sons.

When I wasn't watching Frenchy, I dwelt upon the news of War Bag's abandoned family. I could not make up my mind about this thing. Where were my convictions? They had fled before the force of personalities greater than my own.

It was easier to pass judgement on the unfamiliar names in a newspaper. But I was no longer removed from life, listening with objective disinterest like a magistrate. Now I sat in the witness stand, perspiring beneath the expectant gaze of my fellow conspirators, whose eyes seemed to say, "judge not, lest ye be judged."

So I didn't judge War Bag harshly, and I can make no other rational explanation for the lack of passion I felt at the news of his great failing. He was a man prone to conflicts I could not understand, and bound to a coda which was not penned by any democratic body, but by himself alone. He had made a choice, and whether or not I felt it was wrong, it was he who had to live with it. My opinion did not matter. What was done was done, and no amount of outrage on my part or on the part of Monday Loman would change it.

Perhaps my jealousy of the dead boy had a hand in this dubious acceptance of War Bag's discrepancy, but I cannot say for certain. Was I glad he was dead? All I was sure of was that I was not yet ready to return to the aloof, sterile, window world I had once called home.

As though to draw our attention away from his failure, War Bag set himself to the task of making good on his word. It seemed in the weeks that followed Adobe Walls we worked like never before.

We struck a wealth of buffalo herds both large and small as we went deeper into Texas. War Bag's favor in our eyes was restored. Our forgetfulness of his sin grew with every hide that piled up in the bull wagon, another layer of scar piled over an aging wound.

The Frenchman proved a skillful, prolific worker, and I believe our dislike of him combined with his own productivity increased our yield significantly. We worked hard to best him, and though we weren't friendly enough to compare tally books, I am sure that we were matching him hide for hide.

Our attempts at conversation with him were few and mostly ignored. Frenchy was usually fast asleep after consuming a small portion at supper, and went straight to work after an equally modest breakfast.

One particularly hot day he stripped himself to his waist, revealing an intricate tattoo emblazoned on his grimy back. It portrayed in dramatic detail the image of Christ crucified, his bloody, thorned head radiating a sunburst of holy light. I had never before seen a tattoo of such magnitude and visceral beauty. It must have taken hours to complete, and it spoke volumes about the Frenchman's tolerance for pain.

Monday was delighted at this. He must have seen the tattoo as an opportunity for conversation, and a possible reconciliation betwixt the little skinner and all of us.

Over supper, he brought it up to Frenchy.

"I noticed, Frenchy, that you've got a tattoo of Our Lord on your back."

We all watched the Frenchman as he supped his beans and said nothing.

Monday cleared his throat and tried again.

"Are you saved?"

Frenchy snickered bitterly. He gestured to his back as he continued eating, and spoke carelessly around his food.

"That's an old sailor's tattoo, *mon ami*. To turn the whip of the bosun's mate." He laughed. "There are many men who would sooner flay the flesh off their own mother's backside than lash the image of Christ."

Monday paled and went back to his beans.

"You were a sailor?" I asked.

Frenchy turned to me, and it was my instinct to look away. He had a hard stare, and an air of constant challenge about him.

"For a time."

He left nothing further to be said. He finished his supper and went to sleep.

"The boys don't care for that feller," I heard Roam tell War Bag the next day.

"The boys don't boss this outfit," War Bag said. "I do."

"You should'a talked it over with them. We didn't hardly need no other skinner."

"Won't hurt none. We'll need men like him, anyhow."

"What you mean by that?" Roam asked.

"I 'spect that little French cuss has plenty of fight in him."

"Plenty more'n we need, I guess."

We met other runners in the coming weeks, and they were welcome informants on the movements of the Indians. Our experience at Adobe Walls had induced us to think that the Indians would be a constant danger, but really we saw very little of them. Those that our outfit did see came only to beg or trade. The wild ones were nowhere to be found, though we heard news of some instances of bloodshed and skirmish.

Gradually, our fear turned to apathy. Only War Bag and Frenchy had nothing to say on the matter of Indians, as though they were gravely waiting for something.

"Do not talk lightly of death, monsieurs," Frenchy warned us once. "Else she will come heavily to us all."

Among the other hunters we met buyers who made their rounds among the many camps of those who persevered that summer among the threat of scalping. Sometimes they were caravans of men with hired on freighters who emptied our carts and lumbered off to Griffin, Rath City, or Austin to sell the cargo at double the price they'd paid. Other times they remained stationary in camps surrounded by piles of treated hides and dozens of smokehouses, the bounty of other runners, ready for shipment to the towns and ultimately to the railroads and the East.

Wherever we met them, War Bag did business, thus freeing up our wagons for more profit. He reasoned that while we might get a better price taking the hides into town ourselves, we could triple our output by staying in the field and selling to the middlemen. He was right.

The harvest was plentiful. We could not move more than two or three miles before we struck another herd. Sometimes we had as many as three skinning camps set up at a time, with one man set watch over a few hundred hides and three or four great meat vats. We filled our tally books with cramped scratchings and watched the paper and coin passed to War Bag with glowing hearts, knowing the tally books we kept entitled us each to a fat share at the end of the season. War Bag kept the money in a locked box in his saddlebags which we all called 'the poor box.' The bags were always on his person, bulging with the promise of riches to come.

It was during this time that War Bag allowed Jack an advance on his pay so that he could buy an older model Winchester rifle from another outfit and return his borrowed Henry.

Immediately upon bringing the rifle back to camp, Jack sat down with the rifle muzzle on the ground and began to hammer a nail into the butt.

"What in hell are you *doing*, Fats?" Fuke asked him.

"Well," Jack said, banging away until we all thought he would shear the wood, "when I traded off my paw's ol' Hawkens, I pried this heyar coffin nail outta the stock, so's I could pound it into m'own gun oncet I got one."

"Coffin nail?" Monday exclaimed, suspiciously.

"Shore," Jack answered simply.

"Fats," Roam said, giggling. "What in the hell good does a coffin nail do you?"

"Why, a feller with a coffin nail in his gun is shore t'kill whatever he shewts at."

"That's *heathen* nonsense!" Fuke exclaimed. "Fats, are you a white man or a *damn* Indian?"

Fats shrugged. "Allays worked fer my paw."

"Jack," Monday said, a little apprehensive. "Where'd you get that coffin nail?"

"I dunno. It were m'paw's." He thought about it for a moment. "Might be he pried it outta his daddy's coffin."

Fuke rolled his eyes.

"Oh *Lord*. It's a goddamned *pagan* family heirloom."

Frenchy tilted his head back and let out a long, unnerving cackle.

My thoughts visited Billy Tyler in his crowded grave less and less, and War Bag seemed to have left his son behind as well. Though he spoke little, we could tell his mind was not so burdened now. Whatever pain he had felt had faded some. He laughed again at the campfires where we regaled each other with stories, jokes, and songs.

Fat Jack spoke often of his beloved plateau, and the blue ridges that could be seen from the back porch of the cabin where he had been born. The big Missourian stroked his cat and talked fondly in his slow drawl of the feeling of crackling leaves and brown pine needles under the thick pads of his bare feet, and the way he and his 'paw' would catch fish with their hands in the creek that ran near their spread.

He once told us of a harrowing encounter with a young black bear. He had gone off alone to fish in his peculiar manner, and had rounded a bend in the creek only to happen upon the beast stooped in the cold water ahead and engaged in the very same endeavor. The big bear had left the quick little blue gills and loped through the water after Jack, apparently deciding that a barefoot Ozark hillbilly in his prime was a delicacy easier caught and longer savored than any fish.

The chase up the creek through its center had led finally up a tree, across a ravine, and down a steep embankment. Jack and the bear both had lost their footing in their mutual haste and the scene the Missourian described, of the bear tumbling past him even as he himself rolled end over end, still creases my cheeks.

Fat Jack had a whole passel of stories concerning this black bear. It seemed that after this first memorable encounter, the two became lifelong enemies. As most adolescents spend their formative years crossing swords with a schoolyard rival or neighborhood bully, Jack had an irregular series of scrapes and mishaps involving this bear. He told us that before he had departed for the West, he had engaged the very same bear in a bloody contest to determine ultimate supremacy, but that the bear had escaped him, taking his favorite knife along in its shoulder as a parting memento into the woods.

"I used t'think if'n I could kill thet ol' bar, I could do most anythin,'" he said regretfully.

We spoke until the fire glowed low. Fuke laughed, and Monday talked of his dear Paula Ann and his baby daughter back home.

Ninth

I Decided I Wanted To Shoot a buffalo.

Maybe it was the stories I'd heard each night passed back and forth between War Bag and Fuke. There seemed to be a thrill to killing that I wanted to taste. I also felt that after my skinning competition with War Bag, I had earned the right.

"What do you want to shoot for?" Monday asked, when I tried to get his opinion on the matter.

I shrugged.

"It's just something I'd like to try, I think, before the end of the season," I said.

Monday thought it over for a moment.

"You think he'd give you a shooter's share?"

I wanted the experience, not the pay.

"Maybe just for the day I shoot. I don't know that I'd be good enough to do it more than once." I hadn't held a rifle in years, and my Volcanic pistol was as pristine as the day Cutter Sharpes had sold it to me.

Among all the men I was the least skilled, and probably the least paid. The idea that I might increase my earnings if I learned something new reinforced my desire to shoot. I finally made up my mind to ask War Bag.

Of course making a decision and executing it were two different things for me, especially when it came to having to approach War Bag. I was like a youngster afraid to ask an adult to participate in something I might end up being told I was too young for.

One hot day as Jack, Frenchy, and I were pounding stakes through the corners of the big fifty pound hides, War Bag came striding into camp with Roam toting Bullthrower.

"Hey War Bag!" I called up to him, trying to remain as nonchalant as possible, even though I knew if he denied it to me, it would probably assume monumental proportions in my mind.

He stooped over the fire and lifted the pot lid to see what Monday was fixing up.

"What's that?" he grumbled. I wasn't sure if he meant the cornbread or me, but I answered anyway.

"I want a turn at shooting," I said, and I was proud that I had kept my voice steady, even though I felt the weight of his impending reply before it had even left his lips.

"Sure," he said without looking at me, and dug out a piece of cornbread with his hip knife.

I stooped and hammered at the stake to hide my pleasure, and though I didn't see it, I'm sure Roam was smiling too.

I even felt an approving clap on the shoulder from Frenchy.

"It will be good for you," he whispered.

It was three days before we saw another unclaimed herd. The prairie this far south was teaming with buffalo outfits. The booming of the guns during the day was as constant as the songs of the crickets at night. It was hard to think that there was any threat of Indians with so many big fifties about.

We turned southeast in the lazy direction of Ft. Griffin, near the Clear Fork of The Brazos, where we would eventually spend the winter. We forded the Red Forks, and heard more news of scalpings and stands against the Indians. It seemed the boldness of their attacks matched the audacity of the hunters who plunged further into their territory. Six columns of Federal troops had been dispatched to deal with the situation.

"I pity those poor Indians," Monday said thoughtfully.

"Don't be fooled, Sin Buster," said War Bag. "Them Indians been fighting each other for hundreds of years before we showed up. If they ever was to all of 'em at once join up against us, they'd probably beat us all back to Plymouth Rock. But they got their own feuds that not even the threat of losin' their land'll heal."

"Maybe they're banding together right now," I said, thinking of what we'd seen at Adobe Walls. "Maybe they're learning."

"Hell," said War Bag. "Most civilized men ain't even learned that yet."

On the first of September we crossed the shallow Pease and came across a herd of about seventy head impatiently watering at the north bank of the Wichita.

Roam scouted the herd, and returned twenty minutes later with the good news.

"Nobody in sight," Roam called, too anxious to wait until he was back in camp.

"Alright, let's see about these shaggies," War Bag said.

My chest ached as I watched Jack, Frenchy, and Monday set up camp. War Bag and Fuke unlimbered their big rifles.

As though War Bag could read every thought printed on my face, he looked over at me and held out Bullthrower.

"Well, boy?" was all he said.

"Lord, help us! I don't *believe* you're going to let him shoot," Fuke said, checking his own weapon and shaking his head as he did so. "I mean, the son of a bitch is from *Chicago*. Might as well try and teach an *asshole* to piss."

"He'll learn fast enough," War Bag assured Fuke, as I hefted Bullthrower. I was a little intimidated by the rifle. War Bag and Roam had taught me to clean it, and I had done some mock-aiming, but never actually fired it. I swear that learning to shoot on that old beast made every gun afterwards seem like gravy.

War Bag passed me the loads and his forked stick, and patted me on the shoulder.

"Roam here will lend you a hand. I'll be by to check on you."

"Ah . . . I suppose *you'll* be fetching *water* then, will you?" Fuke said to War Bag. He laughed, his rifle over his shoulder lazily.

War Bag didn't smile.

"You watch your tongue, Lousiana, or I'll bust that pretty gun of your's over my knee."

Fuke twiddled his fingers and ambled off with Napoleon.

"I'll take the *west*," he called over his shoulder. "You be sure and put *Juniper* to my south. I don't want him putting a window in my *skull* accidentally."

"It wouldn't be an accident!" I called out to him as he walked away, but he made like he hadn't heard.

Roam took me to a stand of low brush on a slight incline that gave me a decent view of the herd as it watered. We hobbled our horses a few feet behind us, down the incline and out of sight of the herd, and crawled on our bellies up to what would be our vantage point.

"Y'see that ol' heifer standin' a ways off?" Roam whispered to me, pointing at a buffalo cow munching grass plaintively. I could not tell the difference between a bull and a cow at a distance of a few feet let alone the two hundred and some odd yard length Roam had put between us and them.

"I guess so," I said dubiously, raising the rifle to my cheek.

"Now hold on, there," Roam said, putting his hand on the barrel of Bullthrower. "Fust you gots to get into position. Take out that stick the ol' man give you."

I produced the forked stick and Roam took it from me and drove it into the earth so that the crotch rested about thirty inches from the ground.

"Go on an' lay that ol' cannon in there."

I rested Bullthrower in the crotch of the stick. This necessitated my sitting, instead of lying on my belly to shoot.

Roam crouched beside me.

"Now take a look in that glass and center on that ol cow's hind end."

"Shouldn't I aim for the head?"

"Naw, they skull's so thick, sometime the bullet don't go through."

From across the herd, there was the boom of Fuke's rifle, and we saw the puff of smoke.

"That'd be Fuke," Roam said.

I looked in amazement. The buffalo on the far side of the herd began to mill about, but near us there was no hint of alarm.

"Why don't they run?"

"It's herd instinct," Roam said. "What you got to do is bleed the leader. Them other buffs'll get agitated. They'll prod her a bit, tryin' t'figger out what's wrong. While they doin' that, you pick 'em off one at a time. Whichever one starts to bolt. The leader's the only ones they care about."

I lowered my head to peer through the scope. Through the little tunnel of vision, the buffalo down at the river were as clear and as close as Roam. I could make out a calf I had not spied before suckling its mother. I swivelled Bullthrower in the rest sticks, and found the old cow, looking bored.

"What you don't wanna do is run 'em off. Ever' shot after that cow has got t'be a killer. Don't ruffle 'em up, or we'll have to chase 'em."

"How many do I kill?"

"No more'n what your skinners can handle. I'll tell you when."

I nodded and chambered a round. The forceful action felt assertive. I peeked through the buffalo bone scope again, and moved to cover the lights of the old cow with the axis of the stadia hairs.

"'Member t'aim higher for the scope."

I adjusted accordingly, and took a breath.

"Squeeze," Roam whispered.

I squeezed the trigger and the big gun boomed in my ears, startling me. Roam had warned me of the kick and told me to hold the gun with respect, 'else it wouldn't respect me. I remembered the lesson, and though my grip was sure, I still shuddered with the recoil.

It took a half a second for the smoke to clear. To my dismay, I saw the old cow bounding across the river with her followers in tow.

"You missed her," Roam told me.

"Well, shit," I said.

"Come on," Roam said, shuffling back to the horses.

The buffalo ran for about two minutes before slowing to a trot and then returning to their previous mood of complacency.

We rode alongside, barely within sight. The buffalo had bad eyes and a sense of smell that was sub par at best. They were an animal born to die.

As we rode alongside them, I noticed Fuke atop Napoleon following, too, on the far side. I cringed at the thought of what epithets he was most likely devising for me.

We took a position at the same distance overlooking the herd from a higher hill than before.

Soon we found the cow again, and I set up Bullthrower.

The sun was shining down from a blue sky, and it was hot in the dry grass with no shade other than my hat. The gun metal was warm to the touch. It was a fine day to die or be killed. As fate would have it, it was the cow's.

I watched her for a few moments through the glass. She had escaped death momentarily and stood in the field with no memory of having done so, and so no fear. Her skin hung low from beneath her woolly head, and her tail flicked nonchalantly at her hind legs. Her working jaw pulped the dry grass. In the background, her disciples lazed about.

I curled my finger around the trigger and squeezed. The recoil jerked my eye out of the scope. At first I thought I had missed again, but then the old cow sat down suddenly, like a dog swatted on the bottom. She regained her feet, slid, and looked around.

"Quick, now. Again." Roam said.

I worked Bullthrower and peered again through the glass. There was the cow, looking confused, her eyes rolling. A dark wetness appeared on her right rear flank. A young bull trotted up and nosed at her.

I centered the crosshairs on the bull's head, and fired again.

The bull spun about as though it had suffered a right hook from Sullivan, then recovered and trotted off, shaking its great head.

"Don't aim for the head, I told you," Roam admonished. "Hold on the neck. Go on."

Fuke had begun firing across the way. I chambered a third round and centered on the neck of another bull that had come up to inspect the stumbling cow.

The bull dropped hard. It took me a while to familiarize myself with the delay that occurred while the bullet sought its mark. It was like tossing up a stone into a high arc and watching it float before finally crashing into a pool of water. I had time to raise my head and watch the bullet strike, if I aimed at a target far enough away.

My fourth shot missed, and my fifth caught a fleeing cow through the ribs as the herd thundered off again. The cow gradually fell behind, then sat down in the wake of her fellows and watched them go.

"Not too bad," Roam said. "But keep 'em from runnin.' Once you take a shot, move onto the next one."

I said nothing, but looked at the cow, through the scope. She sat there, her great flanks heaving, then fell to her side.

I aimed for her neck and blasted her dead.

"Try not to shoot a buff twice," Roam scolded.

"She was in pain," I said.

Roam looked at me for a long time, his dark face unreadable. He shrugged.

"Well," Roam said. He put his hand on my shoulder. "Jest try."

On my next attempt, War Bag joined us at our position. He watched me cut the herd as Roam had taught, and this time they did not run. I had what was called a 'stand.' The wounded cow stumbled about, and I picked her attendants off like a sharpshooter, until she bled to death. War Bag said nothing, only grunted in satisfaction as the buffalo died at my hand one after the other. He left a bucket of water for us, and refilled our canteens.

Roam dumped some of the bucket on Bullthrower, and the barrel hissed angrily, releasing a little cloud of steam.

"We can rest up now," Roam said. "Let Fuke finish 'em."

I put Bullthrower aside and lay down on my back, looking up into the sky. The sun had only just begun its descent. Roam propped himself on his elbows and chewed a dry blade of buffalo grass.

We listened as Fuke's gun boomed twice more and then was silent.

My hands felt raw from shooting, and I was stiff from holding the same position, but it beat the aches and filth that came with skinning. I felt the grass tickling the back of my neck and shaded my eyes with my hat. My arms were red with sunburn, and the skin on the back of my neck felt hot.

My eyes wandered among the cloudless skies before falling on Roam and the pistol at his side. I thought of the day so long ago when I had cringed to share food or a canteen with him.

I also thought of Roam's ex-comrade, Arlo Flood, covered with ants and beads and pilfered sugar, and Frenchy, grabbing at his pistol.

"What?" Roam said. His eyes had come to regard me, and found me staring.

"Would you have killed Frenchy that day?" I asked.

Roam knew right away what day I was talking about.

"If he didn't shoot me fust."

I was quiet, remembering how I could have taken my own gun out and interceded. Why hadn't I? The Frenchman hadn't even been facing me. I'd been in no danger.

"Well, it would've been self defense, I guess," I said.

Roam shook his head.

"Weren't no judge around to say that. 'White man get plenty of friends all of a sudden, if'n it's a black man kills him."

I was quiet.

"I could've stopped him. I had my gun."

Roam looked at me, and chewed his grass blade.

"Well," he said, "pointin' a gun at a buffalo and pointin' a gun at a man ain't the same thing, I 'spect."

"No," I said. Was that it? Was I unable to kill a man?

From the direction of the field, Fuke's Winchester cracked three times in succession and we got to our feet in alarm.

Roam had already taken up his Spencer when we realized Fuke was just running the twelve or so remaining buffalo with his horse for fun.

"He goin' get hisself kilt one of these days," Roam muttered.

Below, Fuke gave chase across the killing field. The carcasses of all those we'd killed that day provided a natural obstacle course. He steered Napoleon expertly through the maze with only his knees, firing his Winchester at the fleeing survivors, who seemed to be mostly old bulls.

"He sure can ride," I remarked.

Roam said nothing.

To the north, Jack and the bull wagon came upon the latest kills. I saw Frenchy hop down and make for a carcass, descending on it like a buzzard with his steel. Soon it was shining red in the sun. Somewhere out of sight, Monday was staking out the hides. I thought it was a bad idea to leave Monday behind, unarmed as he was. I said so.

"You right about that," Roam conceded. "Oughta figure out a rotation."

I watched Fuke's rifle belch angry smoke as he rode down a tawny bull. The creature tumbled in mid run and skidded to a stop in the grass.

Fuke let out a whoop and urged Napoleon toward the others. These were retreating at a gallop to the south.

"Why doesn't he just let them go?" I wondered aloud.

Roam groaned, rising to his feet and stretching. He looked off to the north.

"Here come Monday and War Bag."

To the north, War Bag was loping up on his black horse and Monday's team of mules marched behind. As I watched, Monday brought the mules to a stop and climbed down from the wagon, ready to make camp.

Fuke's rifle sounded twice again. He was empty, but he still rode at the hind legs of the last buffalo, running them for the sheer thrill of it.

Then one of the bulls in the rear stopped and turned abruptly into Fuke's charge. It leveled its head to the earth and brought it up in a violent jerk. I saw the red flash as the old bull hooked Napoleon near the back legs and the Appaloosa's guts spilled over the beast's shoulders. The horse screamed and reared, flinging Fuke right off its back into the grass.

Roam plucked Bullthrower from my hands and leveled it down at the killing field. In the span of three heartbeats he blasted the old bull twice, wounding it with the first, and sending it toppling to earth with the second. It lay on the grass, a kicking mound of muscle and hooves.

Napoleon jumped up from the dying bull and cantered off, shrieking the whole way. It bucked and shook its mane, trying to rid itself of what must have been unbearable pain.

Fuke stirred in the grass a few feet away. He slowly rose up on one elbow, hatless.

"Come on," Roam said from behind me. He leapt up on Crawfish, still holding Bullthrower in one arm. He tossed me Othello's reins, and with a click of his tongue, went plunging down the incline.

War Bag and I arrived at about the same time. Roam was helping Fuke to his feet, and urged him to try standing on both legs before walking. Fuke jerked away from him.

"I'm alright, *dammit!*" he growled. Then he saw his horse, his beautiful Appaloosa, standing nearby.

Napoleon's underside was a mess of blood. In its mad run, the horse had tangled up its legs in its own intestines, and stood grotesquely hobbled and wild-eyed, stomping its feet and screaming. The tendons on its long neck stood out like taut steel chords, and its whole body was bathed in foam.

"Lord," Fuke muttered.

War Bag and I dismounted. The bull suddenly gave a snort, and Roam shot it with the Spencer three times. It did not stir again.

Fuke found his rifle a few feet away. Walking toward it brought him closer to Napoleon, and at his approach the horse tried to take a step and stumbled.

Fuke held out his hand.

"Easy. Easy."

The horse shook its head and snorted, eyes rolling wildly.

Fuke slid a cartridge out of his belt and pushed it into his Winchester.

The horse began to buck again, and the gash in its undercarriage tore wider. The sound was wet.

I cringed. The scene was horrible, but impossible to shut out.

Fuke raised the rifle quickly to his cheek with his characteristic flip and shot his horse through the head. It dropped like a puppet with cut strings, or a bag of bones.

"Buff tea," muttered War Bag.

"What?" I said.

Fuke stood over Napoleon.

War Bag took Bullthrower from Roam and walked back to his horse and mounted up.

"What did he say?" I asked Roam.

"Buff tea," whispered Roam. "Fuke just took a big ol' drink."

Tenth

The Weather Turned quite suddenly one morning.

It was fine climate for drying hides, but not for men. The summer heat beat on our backs like the breath of a tired dog. Fuke was of a sour disposition for a few days after losing Napoleon. He repeatedly offered to buy our horses from us, but nobody wanted to ride shotgun in the bull wagon with Jack anymore than he did.

Boredom overtook us, and there was little to do after we had finished our work but sit under the wagons and watch hides tan and meat cure.

Insects flitted through the dry grass and dropped dead when they got too close to the arsenic. This was an endless source of amusement for Frenchy, but did not prove very engaging for the rest of us. It seemed that the time to pack up camp and move on could not come fast enough.

A week passed and we saw no more buffalo, nor any sign that they had been south of the Wichita Forks. There had been talk of turning back north, or west. War Bag's argument was that there was little sense in going over the same ground. Roam was for going back, but I think it had more to do with his chronic unease about Texas than anything else.

We awoke one morning to find Jack unpacking his rain gear, though the sky was unclouded and bright.

"Whisper licked his fur agin the grain," he explained. "So I 'spect a gullywasher."

"*Redneck hocus-pocus*," Fuke told Jack sleepily. He rose and kicked at the three-legged cat out of spite.

But by noon clouds were drifting in from the northwest, and a cool wind ruffled the grass. It would be the first real rain we had seen all summer. There had been overcast days, but the heavy clouds had always passed over and dropped their burden elsewhere. This time it would be dead on.

It turned out to be a real frog-choker. The land and the sky went gray and old with it, and we were soaked to the toes of our boots before we could scurry for our rain gear. Roam found his tunic, Fuke his capote, and the rest of us donned buffalo coats (all save Fat Jack, who smiled and said nothing, the water running off his oil coat). It was a hard rain, and the sound of every drop striking the earth rolled over the land like an ovation. The ground turned to mud, and the going got slow and hard.

By three o'clock the tempest died down to a light sprinkle that would have been pleasing had we not already been drenched. There was a peaceful stillness over all the faded landscape. The animals shook the water from their bristling flanks. On days like these back home I would walk along the lake shore with the collar of my topcoat turned up, and watch the thousands of tiny drops erupt on the surface of the water.

"It's proof of God," Monday told us. His face was very white against the drab sky.

"What?" Roam asked.

"The Lord, renewin' the land. If you've ever leaned in the doorway of a farmhouse and watched the rain turn the earth to chili . . . seen the leaves of the green beans dance, and smelled that . . . I don't know . . . fertile smell in the air. It's proof that He's there, and that He cares."

"For being such a *pulpiteer,* how'd you end up with that pagan name -*Monday*?" Fuke asked.

The muleskinner shrugged.

"My paw wasn't very religious," he said. "My maw told me she fought him tooth and nail. I was supposed to be named Michael, but paw said he knew too many Michaels of ill temperament."

"Were you born on a Monday?" I asked.

Monday shook his head.

"It was a Sunday," he answered.

"No doubt you were dropped in a *pew* and reached for the *hymnal* before the nip," Fuke said, chuckling.

Monday blushed.

"My paw, he used to drop my maw and me off at church and then wait for us outside. I would always see him through the window, smoking and watching the road. He was a strange man. I used to think he was bad, or he had done something so bad he couldn't go into church anymore. Like . . . maybe God had cursed him for something, and if he went in, he'd burn up. I remember asking him once when I was very small how come he didn't come to church with maw and me."

"What'd he say?"

Monday sighed.

"I don't recall the answer. Just the asking."

"Well what was your father's name?"

"Zachary."

I pulled a blanket from my saddlebags and wrapped myself in it. My nose was red and cold, and I shivered in the saddle. I found Stillman Cruther's red wool muffler and tied it over my face. That helped some, but then my nose began to run.

Winter had given Fall a jump and our knuckles trembled as they gripped the wet reins. The wind picked up and whipped about our legs.

"Still think this is the good Lord's *work*, Monday?" Fuke muttered. He had taken to riding with the muleskinner, saying Scripture talk was a sight better than listening to Jack go on about his queer superstitions.

Monday did not answer. His mules out front were troubled, braying and shaking their heads in the harness. They had not made a sound at the approach of the storm, yet now in this chill wind they seemed tense. He spoke to them, too low for anyone with short ears to hear.

I craned my neck up, feeling the rain on my face. A flock of geese were cutting madly across the murky sky, buffeted by the wind. Then I saw something odd that I never will forget. The entire sky lit up with a crazy, twisting chain of lightning. It flashed out like a bullwhip and in an instant struck in the midst of the flock. They were burned on my cornea, little white 'ems' silhouetted against a purple flash, as of a photographer's powder. There was a weird honking cry and a tremendous crash of thunder. Then twelve or fifteen of them dropped lifeless and blackened from the sky into the wet grass all around us like great, feathered hailstones.

My mouth fell wide open.

"Great God! Did you see that?"

Fuke was the first to laugh.

He fairly leapt from the wagon seat and stumbled into the swampy grass where two dead geese lay smoking. The smell was an acrid mixture of rain, static, and burnt meat. Fuke gingerly reached out and grabbed them by the necks, withdrawing his hand quickly, unsure. Then he snatched them up with aplomb. He lifted one in each fist and stood smiling.

"There's proof of God for you, *Sin Buster*! Manna from *heaven!*"

We all laughed, exhilarated by the unnatural occurrence and warm with the knowledge of a couple of cooked goose dinners for the coming week.

Jack did not seem so happy, though, and shook his head.

"Y'all ought t'leave them geese be."

Fuke rolled his eyes as he returned to the mule wagon with the two dead geese.

"Oh come *on,* Fats! Don't tell me your three tittied backwoods witches got anything to say about *this*?"

Jack scratched his head gravely.

"No, only . . ."

Fuke cut him off.

"Well I'll be *damned* rather than look this gift hoss in the *mouth.*" He plopped the two fat birds up into the wagon bed.

We paused and gathered up what geese were worth it into the camp wagon. Monday agreed to sit in the back and pluck them if Fuke would take the reins for awhile.

Fuke assented, but his command of Monday's mules proved less than masterful, and they soon fell behind. We could hear him cursing the animals through the rain. Gradually he grew hoarse or tired. I fell back to keep an eye on them, and rode in their tracks. A little trail of blackened feathers began to flit from the back of the wagon and float between the ruts, as Monday went to work. I frowned at the sight of them, for I was reminded of the turkey feathers we'd seen outside the pumpkin rollers' camp.

The chill wind died out. The rain continued on for another hour, and we dozed in our saddles. Jack sang a low song as he drove the bulls on, and the creaking of the wheels and the rocking motion of Othello grew hypnotic. I tried to make out Jack's words, but the melody was inseparable from the lyrics. My eyes were heavy. I flinched awake several times before giving up the battle and slouching in as comfortable a manner as I could muster. I slept. Jack's wordless singing was the last thing I heard.

It was one of those naps that seem to take place in an instant. When I snapped awake, Jack's singing had stopped. The rain was gone. Further, Othello had stopped to crop the wet grass. Shaking myself awake, I saw that there was no one in sight.

The gray prairie stretched out empty all around me.

I had heard the phrase *lost without a trace,* but never truly understood the meaning of the words. I thought it was reserved for the snowblind and those unfortunates who fell overboard at sea. Yet here I was, as lost without a trace as a man could be. I had fallen behind and no doubt my comrades had continued on unawares. I thought to resume my traveling with a nudge to Othello, but who knew if the horse had strayed from his course as I slept? There were no tracks to follow (not that I could follow them anyway), no easily spotted wagon ruts. All around me was the empty gray stillness of the rain-soaked prairie, a boundless, gate-less Purgatory.

I remembered Roam's advice not to go looking, but I saw no evidence of the wagons. That terrified me. I turned in my saddle.

There in the grass were the almost imperceptible tracks of Othello. Would Roam be able to find them? Perhaps my absence had not even been noticed yet! How long had I been asleep? I could see mosquitos flitting up from their grassy shelters. The hair on the back of my neck prickled. I couldn't very well just sit here until night came.

I thought of Roam's advice about firing a rifle into the air. I had my Volcanic pistol. In the storm I would have had no chance to be heard, but in this stillness, I found a hope and grabbed it. I fished under my coat and prayed that the powder wasn't wet. I pulled back the hammer, pointed the pistol skyward, and squeezed the trigger.

I was almost startled by the ensuing shots. I had not truly believed until then that the gun would work. I lowered the pistol with new respect. It was a thing now alive in my hands, its acrid breath dissolving in the cool air. I waited.

I was ecstatic to hear in the distance (from which direction I could not readily ascertain), the reports of a rifle in answer. I had not slept so long nor strayed so far as I had feared! It seemed to me the shots had come from nearby.

I raised my Volcanic again and fired, unable to contain the smile on my face. In a few moments there was another answering shot, closer, and off to my left.

I turned Othello to face that direction and stood in the saddle to see. There was a low dip in the land about a hundred yards out. Then there was another shot, and I saw the smoke flitting in the air.

I put my gun away and pulled my muffler down around my neck. Cupping my wrinkled hands out over my mouth, I shouted,

"Hey! Over here!"

Roam came up over the rise. Though it was hard to make him out, I recognized his dark skin, his spotted piebald, and his union blue coat. As he appeared, he fired another shot.

I waved my arms happily at him, grateful to have been found. I was still advertising myself like a fool when a bullet creased my right cheek. It had sounded like a fly in my ear, and I had mistook the sharp pain for a mosquito bite. I slapped my hand to the cut, and when it came away, the palm was red with my own blood. As I pondered the significance of this, another bullet struck the earth beside Othello with a wet plop.

With a revelatory tremor, I realized that the black man on the piebald was not Roam Welty.

Eleventh

My Whole Body Tensed up as I came to understand that the man on the ridge was firing not to signal me, but to hit me. Othello must have sensed this change in manner as animals do, for he turned about in place, wanting to go.

I kicked him. It was an action that summoned up all the energy I had, for the knowledge that I had not only been fired upon but actually shot, had blanketed my body and senses in an all-encompassing numbness. Othello took off in a leap, with me bouncing on his back like a loose kettle in a runaway buckboard. My fickle hat left my head, seduced by a persuasive bullet never to return.

The chill wind broke across my face and I clutched the reins like lifelines. Behind me, the rifle popped again. The sleepiness of my limbs was replaced with a nervous, kinetic vitality.

The man on the piebald gave chase. Though he was far behind, his mount was a wild charger and he gained rapidly.

There was no cover. We tore across the wide expanse, a train of noise and gun smoke. I knew that if one of his bullets so much as nicked me I would likely fall from the speeding horse and break my neck.

For all my fear of Indians, I was now at the mercy of a 'civilized' killer. An outlaw, or some other blackguard. I considered the Union coat. Was he a soldier? Was this a misunderstanding? Perhaps if I stopped and surrendered, I would be able to explain? Another shot, and I distinctly heard the crack as it whipped by my ear very close. There would be no explaining.

As my pursuer gained ground, I saw that he was indeed a Negro, but lighter in hue than Roam, and larger. It was foolish mistaking him for Roam. The pattern of his horse was all wrong. His hat slipped off and filled with air behind him, the rawhide cord tight against his neck. The reins of his horse were clamped between his teeth, and he stood in the saddle and levered a Henry rifle. I gave no thought to returning his fire. It took all my concentration to keep astride the rocketing horse.

The chase was relentless, and as the froth from Othello's coat wet my legs, I knew no hope. My horse was near his limit, and would collapse at any moment.

Then we struck a mass of holes and earth mounds. It was a prairie dog town. I was terrified that exhausted as Othello was, he would stagger and throw me, perhaps laming himself in one of the myriad of burrows, or crushing me beneath his weight. Instead, as though God had heard my thoughts, the piebald behind me gave a cry and went down head over heels. What miracle insured that my pursuer should tumble and Othello should not, I do not know to this day. I welcomed it then, and still give thanks for it now.

I looked over my shoulder and saw his horse roll over him, and prayed he had broken his skull. When I turned back to the land before me, there came the cries of several horses, and the exclamations of men's voices, followed by a brief rush of color beneath me as Othello leapt over a line of Indians laying with their horses behind a low ridge.

Othello topped the rise and staggered down the grade. There were the wagons, the blessed, familiar wagons plodding along sleepily. I spied Fuke's bright coat and the genuine Crawfish at the head of the train with the real Roam astride him. I heard War Bag's thunderous voice shout something that was lost among the great Indian cry that resounded in my ears.

Looking once more over my shoulder as I made for the wagons, I saw a dozen Indians and their ponies rise up from their position in the grass. I had ridden through the nest of them even as they tailed the outfit, spoiling what would have been a devastating ambush.

The boys knew there was no outrunning them, and Fuke flipped his rifle to his cheek and stood in the seat of the camp wagon, firing over my head. The others took his cue, and I saw Jack take up his new Winchester with the coffin nail in the stock. Roam and War Bag swung themselves with bravado out of their saddles and jerked their horses flat on the earth to form breastworks. Monday still had a half-plucked goose in his lap as he scrambled for cover.

When I was among them again, Frenchy jerked me out of my saddle and pulled Othello down beside his own horse. I fell to the grass, splattered with mud. Frenchy shoved me across Othello's neck to keep him still. He produced his pistol and commenced to firing.

They were swiftly upon us, whooping and screaming in tones that made the hairs on my neck and arms stand at attention. A blurred rush of painted men and horseflesh rained down like a living storm. A feathered arrow shaft sprung from the neck of Frenchy's horse and the animal gurgled and was still.

Roam rolled with an Indian. I leveled my pistol across Frenchy's back and aimed, but before I could take the shot a great horse interposed itself between us.

I looked up to see a tall, rough hewn warrior on a large gray pony with a rifle across his saddle and a feathered lance poised to strike in his uplifted arm. His long gray-black hair was flattened by the rain to his broad shoulders, and he wore a blazing red paisley shirt with a bone necklace and buckskin leggings. The left side of his face was pitch black. Black teeth shone through a dark hole in that sable cheek. His mad eyes were set in a startling mask of misshapen red, barely passable as a human face. He let out a keening yip and flung his lance down. There was a sound like a sack of potatoes being pierced with a spade, and Frenchy groaned.

I fired up at the nightmarish Indian, but when the smoke cleared, he and his horse had moved on. I saw him bearing down on War Bag. I drew a bead again, but the Indian left his saddle and pounced like a mountain lion. I saw only the old man's boots as he and the Indian tumbled down behind Solomon. The black horse immediately sprang up, again spoiling my aim.

There was a fantastic crash and the mules took to braying. The camp wagon had overturned, and the animals were struggling to tear themselves out of the twisted harness. I saw Fuke tumble out, still gripping his rifle and shooting wildly. A wiry, bare chested Indian disappeared behind the wagon, a silvery knife flashing in his fist. Then a hand clenched my leg.

It was Frenchy. The lance protruded from his side, pinning him to the earth and sticking straight up into the cloudy sky.

After cursing in a rapid volley of French, he screamed at me,

"Take it out!"

All around me was the snap of gunfire, the whistle of arrows, and the screams of men and horses. The rain begin to fall once again. I dropped my pistol and, clutching the Indian lance with both hands, drew it up quickly out of the Frenchman. With the lance head came an eruption of blood and a rip of fabric. Frenchy rolled on his back, propped his head against his horse, and resumed firing.

I took up my pistol with both trembling hands and sent a few shots across Othello's belly, not sure if I was hitting the enemy, and praying I was not shooting any friend.

Those that had not waded into our midst to kill us barehanded were orbiting us on horseback like buckskin comets, stinging at us with arrows or rifles. I do not know how they missed their own comrades, as we were a tangle of men and animals. I saw some of Monday's mules go down bristling with shafts like strange quilled animals, and heard their terrible, dying brays.

My pistol emptied and I flushed, for I was not certain how to reload it. I ducked down and fumbled through my pockets.

Beside me, Frenchy lay still on his side, the rain drops bouncing off his face. He had passed out, or was dead. I glimpsed the wondrous tattoo of Christ beneath his torn shirt. It was marred by the bloody hole where he had been pierced, as though by the centurion.

Somewhere, Monday shrieked. Fuke knelt beside a busted wagon wheel and fired his rifle. Roam stood up and his pistol coughed smoke, the body of the Indian who had jumped him lifeless in the grass at his feet. Jack loomed up like a red coated bear from where he had gone down, his unkempt black hair and beard plastered to his skull. The big Oriental knife was naked in the air, its edge molting drops of blood as it bore down like a harvester's sickle.

I left Frenchy and crawled to where Fuke was crouched. I called to him, but he didn't hear me.

I felt a violent, painful tug of my hair then, and I was wrenched flat on my back. A tremendous weight fell on my stomach. Blinking back the rain, I saw the dark outline of an Indian silhouetted against the washed out sky. He bent towards me, knife arcing down. I caught his wrist and we were frozen, struggling desperately back and forth.

His skin was greasy and hard to hold. He slipped from my grip, but I twisted my body and the blade plunged into the mud beside me. I turned over and clawed to get away, but his iron-like knees locked about my hips. I could not get but a few inches.

My hand found something laying in the grass, and I grabbed it and swung it to ward off my assailant. It was Monday's coffee pot, loosed in the fall of the wagon. I struck the Indian twice in the face and he fell back, the pot freshly dented in my fist.

Then I remembered the knife in my belt. I took it out and barely had time to brandish it before the Indian fell heavily forward again. He gave a low sigh. Where there had been rainy air on my hand that held the knife aloft, I now felt the pressure of his flesh. The knife had gone in him to the hilt. He promptly unrolled from a wild, tense bundle. He laid his head on my breast and I heard him hiss like a strangled snake.

I blinked away the rain. Warmth spread from his belly over my whole hand. I felt sick. I watched the rain strike the Indian's bare back and turn to beads which burst and retreated down along his dark skin. A gray feather from his braided hair stuck to his shoulder. I was too afraid to push him off of me. Not for fear that I would be killed, but that I would disturb his death. He did not sing any song, nor whisper any valediction or curse me in his alien tongue. He coughed once, and was through.

I laid my head down on the grass and stared up into the churning gray sky. I felt unreal, as though it were me taking leave of my body, and not the Indian. I felt his blood spread across my torso. The gunshots and cries seemed so far away. I do not know if tears fell from my eyes, or if it was the rain.

Gradually I shook off the feeling and slid out from under the dead man, careful not to turn his face to the sky. I put my back to the underbelly of the camp wagon, behind Fuke, who had not noticed my plight. The dead Indian lay face down in the grass. I would not look at him. I left my knife in his belly. I had killed a man.

I should write at this point that this is not the murder I alluded to at the start of my narrative. This was self-preservation, not the deliberate act of which I have yet to tell. Still, it is no less a sin. Even after thirty years, some rainy nights I still feel the weight of that Indian upon me, and I wrestle the bed sheets out of naked terror in that instant before I awake.

Fuke flattened himself against the wagon again and dug in his pockets for cartridges. When he saw me, he looked grave.

"Lord. Are you *shot?*" he asked.

I shook my head. There was blood all over my shirt, and my hand was red with it.

"No, it's not mine."

I saw five Indians on horseback riding away. I did not see the big horse with the tall, black-faced war chief.

"I think they're *departing*," he said. He cupped his hands and yelled after them, "Go *on,* y'sumbitches!"

He slapped my shoulder.

"I got *three* of them bastards!"

There was a moment when only the pouring of the rain was heard. Then Roam's voice called out from somewhere.

"Everybody alright?"

Fuke shook my shoulder and answered,

"*Juniper* and me are alive!"

He looked at me.

"Where's that Frenchman?"
I nodded to the horses where I'd left him.
Frenchy looked like he was dozing lazily against his horse.
"I think he's dead," I said.
"Ah'm heyar!" Jack said from the hide wagon, and I saw his great arm pop up from behind the spilt stack of hides and wave his battered hat in the air. An arrow transfixed the meaty part of his forearm, but by his chipper tone, he hadn't noticed it yet.
"Yeah," I heard War Bag say, a little hoarsely.
"We *lost* the Frenchman!" Fuke said.
Roam called out,
"Monday? Hey, Monday!"
Then I heard our muleskinner's voice murmuring lowly. He was behind us, separated from view by the overturned wagon.
"He's over here!" I shouted.
Fuke and I stood up and went around the wagon. The mules in the harness were dead or sorely in need of death. Only Michael was unharmed. He shook the water from his mane as we passed him.
Monday was sitting with his back against the wagon, its contents scattered about him. The elk antlers he had mounted were broken to pieces. The Dutch oven was lying in the grass. A few of the blackened, plucked geese lay around him.
He was crying openly, and trembling. Between his sobs, he whispered a prayer. An arrow shaft protruded from each of his legs, and there was blood running down his face.
We moved as one to help him. I saw that there was an arrow in his neck almost to the feathers, pinning him to the board behind. He had been scalped. The blood poured from the circular patch of exposed tissue on the corona of his head. He was fish belly pale.
"Sonofabitch!" Roam said, when he saw Monday.
We all hurried to free Monday. He just trembled, his prayer interrupted by a momentary mewling of pain that sounded sick and pitiful through his shivering lips.
We cut his legs free, but Roam warned us not to fool with his throat. He pulled us aside and whispered.
"Free up that arrow, an' he might could die."
"He's going to die *anyway*," Fuke hissed.
Thunder rolled across the sky and the rain poured down anew, causing the flow of Monday's blood to fade and renew.
"Our Father . . . ," I heard him say in a weak voice. Each time he swallowed, he clenched his eyes shut.

Fuke took out his knife and leaned forward to cut the arrow from his neck. He paused, wiped his hand across his face, and then shook his head and sat back on his heels. His face was drawn beneath his mustache.

"Nossir," was all he said.

Roam took the knife from Fuke and pushed past him. He leaned forward, peering behind Monday's head. The arrow emerged from the back of Monday's neck and had burrowed its head in the wood. Roam reached in. As he gingerly took the shaft between the two fingers of his free hand, he jostled the muleskinner's head. Monday groaned.

Roam closed his eyes and cut the shaft with a snap.

Monday's shoulders relaxed. He still breathed, but he was shivering.

Roam knelt before him. Touching Monday's forehead with the flat of his palm, he gripped the feathered end of the arrow with his other hand and jerked it out.

Monday winced, but did not cry out again. Immediately a river of blood began to gush from the wound.

He opened his mouth, and blood ran from the corners of his pale lips.

Roam reached around and cradled him like an infant, pressing his bandanna to the wound. The fabric darkened almost immediately.

"Aw," said Monday, and more blood ran out. "Aw. Hold me close."

He shivered and wept some more. Roam encircled the muleskinner with his arms. We saw the blood stain the front of Monday's shirt.

"Our . . . . Aw. Aw. Our Father . . . ."

I touched Monday's hand, not sure if he could feel it. In response he gripped my own tightly. His eyes moved all around, desperate for their last looks at the mortal world.

"We're gonna take your share to Haskell County," I said numbly, though I knew less than nothing about Haskell County or where his farm might be. He gripped my hand tighter, unfocused eyes looking around wildly for the source of the promise. "To Paula Ann."

"We shore will," said Roam, his voice muffled by Monday's shoulder.

"Aw. Our Father . . . . Our Fa-Aw. Aw."

Jack, had been standing at Monday's feet the whole time, his bloodied knife hanging limp in one giant hand, and the arrow sticking through his hairy arm. He began to speak in a quiet voice.

"Are fodder who art in hay-vin. Hallow'd be thigh name. Give us this day are daily bray-ed, an' blest are them whut trass-pass agin us."

Fuke took off his hat. His eyes were shining as he mouthed the words.

"Aw," said Monday, and his jaw was a mess of dark blood. He coughed, and shut his watery eyes tight. His lips wrinkled up and a childish whine came from the corners of his mouth. He shook a little and his nostrils flared rapidly, like those of a scared animal.

His hand squeezed my own until it hurt.

"It'll be alright," I said doubtfully. That made me well up, for I had no sure comfort to offer, and no idea if the last words he would hear on earth were a lie.

Fuke took Monday's other hand, and I saw the tendons of the muleskinner's wrists grow taut, his knuckles white.

"Aw. Holmeeclose. Aw. Holmeeclose. Aw. Aw. Awwww . . . ."

Monday's last prolonged breath hissed out, and whatever induced him to grip my hand left with a shudder.

I shivered, as if his ghost had passed through me in its departure. I gave Monday's cold hand a parting squeeze and cautiously freed myself from his clutch.

Roam eased him gently to the ground and passed one dark hand over his staring eyes, shutting them forever.

Jack got a blanket from the wagon and swept it over him. It floated as it fell, and then took on his shape.

We all knelt around him silently, until our immediate sorrow had subsided enough for us to rise and go.

War Bag was not among us, and we had to see about him.

Twelfth

War Bag's Wounds were terrible. We found him lying in the mud, his coat strewn about him like the trimmed robe of a fallen monarch. The gray pony that had borne the tall black-faced warrior who had lanced Frenchy was there beside him, dead from a bullet wound. The Indian rider was not to be found. Solomon stood protectively over his master, the reins still looped around the old man's elbow.

War Bag was splashed with blood like a spent berserker. The bodies of two Indians lay shot or stabbed on either side of him. He cradled one blue faced brave's head in the crook of his arm like a parent with a dozing child. His beard was flecked with red, and his right eye stared ferally at us as we approached. In the left socket, in place of his eye, there was an arrow shaft which emerged from the front quarter of his temple. At first we could not tell if he was wounded elsewhere. Even the blades of grass around him were bloodstained, as though they too had participated in the fight and now lay wounded.

When Roam rushed to his side, the old man brandished his long red knife and pulled the dead Indian closer. He had to be coaxed into permitting us to approach.

Fuke took off his blanket coat and wrapped it around the old man.

"Didja see him, Trooper? That same one! That same damn one!" War Bag spluttered.

We knelt down around him like subjects around a raving King George.

"What's he goin' on 'bout?" Jack asked. He had finally extracted the Indian arrow from his arm and was wrapping the wound with a length of cloth.

"He's outta his head," Roam whispered, as he looked the old man over.

War Bag gripped my shoulder and wheeled his head about, addressing us each in turn.

"Pierre! Jackie-boy! It's him I tell you, that burnt Dog Soldier!"

"He g'wine t'live?" Jack asked, with real concern. He had absently fished out a square of tobacco and was chewing.

War Bag groaned.

"I think it went clean through," Roam said. "Cain't tell if the rest of all this is his or . . ."

Then I blurted out,

"Lord, he's right!"

"Who?" Roam asked, looking at me.

"There *was* a powder burnt Indian!" I exclaimed.

War Bag nodded, his good eye rolling crazily. "Burnt!" he said. "Same damn one!"

"You seein' things, Stretch," Roam said.

"It's the truth! I saw him." I pointed to the dead Indian horse. "That gray mare...that's his pony."

War Bag said he had seen a powder burned Indian on a gray mare kill Stillman Cruthers.

Roam leaned forward and touched the right front hoof of the dead gray thoughtfully. I thought I saw his dark eyes flicker.

War Bag hissed.

"Where's my goddamn gun?"

He commenced to feeling the bloody grass all around him for a weapon.

"We got to get him in the wagon," said Roam.

"Then what?"

"Make for Griffin. We got to get him a doctor."

"Doc Mooer's dead!" War Bag complained to no one living. "Everybody's dead! Tell 'em to eat the horses, but don't foul the water! Gonna need it...!"

Jack bent and gathered up War Bag in his arms. The old man squirmed and Jack had to duck the arrow protruding from his eye every time he whipped his head about. We convinced War Bag to leave the blue faced Indian's corpse lay. Fuke and I took a hold of the old man's legs and we bore him to the mountainous hide wagon.

"Help me chuck out some of them hides," Roam said.

Fuke and I looked at each other, but Roam clambered up and began untying the high load and heaving the hides and wrapped meat off the top.

I watched them flop down into the grass one by one. Each one represented hard toil and long hours.

"Hey now . . . ," Fuke began to protest.

Roam kept on tossing down hides.

When there was enough room, we took War Bag up and laid him down amidst what hides remained. Roam got up into the wagon.

"Fats, you c'mere and hold the ole man down," he said. "Y'all hunt up whatever you can—guns, hosses, anythin' we can use and carry."

"*Hang* on, now," Fuke said. "Just who in *hell* made you boss?"

Roam stared at Fuke, and pulled out his hip knife.

Fuke and I both took a step back. I remembered the first time I'd seen Roam pull a weapon.

Roam turned the knife around and extended the handle to Fuke.

"That arrow got to come out. You goin' do it?"

Fuke's lip curled, but he went off to scrounge. I paused before following.

"Will they come back?" I asked.

"Losin' braves ain't no little thing to them," Roam said. "Probably they goin' come back for they dead."

There were four dead Indians. Two were those that War Bag had accounted for, and the others had been done in by Roam and Jack respectively. Of the brave I had killed, there was no sign. For a moment I entertained the idea that I hadn't killed him at all.

Atop the hide wagon, War Bag thrashed and Jack leaned over the old man to still him.

"Fuke!" Roam called down. "Get out your fifty and watch for 'em! If they come, they'll be in range of us a'fore we are of them."

Fuke went grudgingly to the camp wagon and found his big Remington case lying among the baggage.

I made my way through the debris, picking up tins of Dupont and whatever else I could find. The camp wagon was a wreck. The front axle had splintered and the right front wheel was disengaged. What articles we had stored in Monday's wagon were scattered and in some cases broken. I still managed to find a good deal that was salvageable, including Frenchy.

The impudent Frenchman cursed me up and down when I bent to search his pockets for ammunition.

I helped him over to the camp wagon. He sat there in the wet grass, his head drooping, his hands pressed to the wound in his side. In the hide wagon War Bag roared, then groaned, and was finally silent.

Roam jumped down and put his bloody hands on his knees. He stooped under the rain, leaning against the bull wagon. He saw Frenchy sitting there and allowed half of a smile.

"You still alive?"

"*Un petit*," he answered. "Come and look at me, black boy, and tell me for how long."

Roam gently lifted the Frenchman's hands, just enough to examine his wound. He tucked his top lip under his bottom and frowned.

"Well, you got a nice hole right through you."

Frenchy's eyelids were heavy, but he looked at Roam. "It does not feel so nice."

"We'll have to lift you, 'less you can stand."

Frenchy made a concerted effort, but groaned throatily and sagged back down, shaking his head.

"No," he said.

Roam motioned for me to help. He looked up at Jack.

"'Thow some more of them hides down, Fats."

"What the hell *for?*" Fuke said from nearby. He was using the overturned camp wagon for a breastworks. The long black barrel of his Remington beaded with raindrops.

Roam didn't look at Fuke, but bent to lift Frenchy.

Of our animals there remained eight of the twelve oxen, Solomon, Crawfish, Michael the mule, and Whisper, who had tucked himself into a corner of the hide wagon during the fight. My own poor horse Othello had been spirited away.

"Two hosses ain't goin' get them to Griffin," said Roam. "We gotta load 'em in the bull wagon. They cain't ride."

"*Hell,*" Fuke muttered. "We won't be able to *afford* a doctor if we're flat busted once we get there."

"We got what the ole man sold already in the poor box. We'll get scalped f'shore if we stay out here."

"Well, I don't know about *you,* but I haven't seen hide nor *hair* of cash since we left Colorado…"

"Look, let's just be off. Why are we arguing about this?" I exclaimed. I couldn't believe Fuke's mercenary attitude.

Roam nodded to me, and slipping our arms under Frenchy's, we hefted him to his feet. The little skinner let out a shriek and shook all over. His head lolled back and the rain struck his bobbing Adam's apple.

We passed the Frenchman up to Jack, who laid him beside War Bag.

"We got to stay close together," Roam said. "Stretch, you and Fuke ride in the wagon and keep real low. Take Bullthrower. We'll tie the ole man's hoss an' that mule to the back. I'll ride out front."

Fuke looked over his shoulder, ready to protest, but he said nothing.

"One last thing," I said, taking Fuke by the elbow and leading him to the mule wagon. "Jack, help us."

Roam looked exasperated, and scanned the horizon. I ignored him. He was not War Bag, and Monday was no stranger.

Jack came over, and the three of us put our hands to the bottom of Monday's wagon and tipped it over his body. At least the scavengers would have to work to get at him. It was little comfort though. Monday would have wanted to be buried. I shuddered to think of his ghost walking the rainy prairie vainly searching for us or for his wife. How many ghosts already tooled about these empty grasslands?

"Do you know where we are?" I asked Roam. I took Bullthrower off of Solomon's saddle and climbed into the hide wagon, careful to step between Frenchy and War Bag.

Roam went over to Crawfish and pulled himself up into the saddle.

"Ought to hit the Wichita if we head southeast. From there it's a straight shot to Kiowa Creek and the Clear Fork Valley."

"How long?" I asked.

"A day. Might be two."

"If those Indians hit us *again*, we won't be able to run," Fuke said, coming over with his rifle on his shoulder.

Fuke and I hunkered down in the wagon, the long barrels of our big fifties poking out over the high lip. We put War Bag's precious saddlebags under the old man's head. He slept mostly, murmuring now and then. Frenchy was pale and staring. Every jolt of the wagon caused him to moan. I pulled the hide up over his face partly to keep him out of the rain, partly to stifle his cries.

Fuke and I waited for the Indian war cry to swell up around us. Jack had given us a bottle of whiskey he'd been saving. We passed it back and forth, filling a reservoir of Dutch courage to lap at the banks of our fear.

War Bag spoke of meaningless things. He spoke to "Pierre" and "Simp," "Sandy" and "Louis." Never once did he ask for a "Billy."

After the first few hours, our minds began to wander. The rain died again to a light drizzle, but the sky remained soupy. Frenchy ceased his cries and managed to sit up slightly in the wagon. He was very pallid, and his eyes were half-lidded.

"How are you?" I asked him.

He managed smile and nodded to me, his lips cracked and pasty.

"I will be fine," he barely whispered.

I offered him some whiskey, but he declined.

The worst was over for him, it seemed. War Bag continued to thrash about, reliving some nightmare, or worse, some past real-life horror.

Fuke aimed across the whole horizon at unseen Indians.

I thought of home and the day I'd set out. How anxious I had been then, and how unwary. I began remembering the slightest details of those first few weeks, and finding the most minute occurrences to be suddenly fraught with foreboding. The theft of my father's watch. The look on the thief's face as he stared dead into the sun from the ruins of his body. Stillman Cruther's scalp in my hands.

Then I saw that Fuke was playfully aiming his rifle at Roam's back.

Fuke must have heard my sharp breath. He turned to look at me, and lowered his weapon a little ashamedly.

Roam's rifle cracked, shattering our senses. We bristled like wary hedgehogs. He had signaled someone.

My heart stood stark still in my breast. Six dark skinned men on horseback were riding toward us. They had long lances propped on their saddles, and their ample hats were tilted down against the cold rain, obscuring their faces. Some of them were wrapped in candy colored medicine bow *serapes*, so bright they were like beings swaddled in rainbows. A seventh man in the rear drove a cart with wheels made from two ponderous tree rounds shod with hammered iron and pinned like great discs to a squeaking axle. Two well-bred horses were tied behind.

One of them broke away from the pack and loped up on a buckskin mare to meet us. He was a Spaniard with a trim beard the color of a cotton cloud.

"*Buenos días*," I heard him say, in a tone that neither gave nor took. He raised a leathery hand in wary greeting.

Fuke watched, a little bleary eyed, and drunk.

"Huh," he said. "Mexicans."

"You speak English?" Roam asked. "Sab-ay Ameree-canno?"

"*No, no—un poquito solamente*," the Mexican answered. "A leetle only."

Roam was resigned to speak in signs with his hands, as I had seen him do with some of the tame Indians that had come begging in the buffalo camps.

"You don't speak...?" I began to ask Fuke.

"Not a *word*," he confirmed. "No need to ask Fats *either*. I'm sometimes suspect of his command of *English*."

Thunder rolled across the sky, and the old Mexican waved his comrades over, calling to them in his own tongue while he smiled a bemused smile,

"*Está bien, muchachos! Son nomas un par de pinche gringos perdedos en la lluvia!*"

There was a chuckle from the other Mexicans and they came bouncing over, the wheels of the cart sounding like mice as they turned, iron tires rattling on stone.

Fuke and I hopped down to meet them.

They smiled congenially enough. The long lances quivering on their saddles were like antique cavalry weapons. What outlaw would rob a man at lance point?

"*Relax,*" Fuke said to me. "These boys are buff runners, same as us."

"He's right," Roam said. "Some of 'em still hunt the old way."

"Can't you make them understand we need help?" I asked.

As if on cue, one of the small band rode up alongside our wagon and peered inside. After a moment, he shouted back to his chief,

"*Mire, patrón!*"

The leader broke off from us and his pony pranced over to the wagon. He stood in his gilded saddle and craned his neck over the lip of the wagon to take a gander at our wounded. When his eyes alighted on them, his face took on a new look.

"*Yo lo conosco. Se llama Tyler.*"

"Did you hear that?" I said to Fuke.

"Hey," said the elder Mexican to us, "Thees *hombre* I know. *Que pasa?*"

"Cheyenne," Roam said, "*indios!* Sa-bay?"

These words had the desired affect on the Mexicans.

"*Si,*" said the old man gravely. "*Comprendo.*"

Then he turned to his companions and barked at them in an excited voice.

"*Escuchan, muchachos! Los malditos Indios están buscando sangre!*"

The Mexicans all erupted into exclamations to themselves, each other, and presumably the Lord above and all His saints. Several crossed themselves. All made ready to depart.

To us, the leader said,

"*Vamonos, mis amigos.* We will go, *si?*"

"*Si,*" said Roam.

Jack nodded gravely.

"Yep. Ah reckin they sa-bayed purdy good."

Thirteenth

The Lancers steered us southeast. We went off in a lope, the Mexicans flanking us like a parade escort.

We rode at the same quick pace for a good hour before we slowed our winded mounts to a walk. Every man kept looking over his shoulder, dreading the shrill scream that would mean the Indians had caught up with us.

Through trial and error we learned our robust benefactor's name. He was Orfeo Largueza, a *'cibolero,'* or buffalo hunter. The men with him were mostly relations of his, and had ridden all the way down the length of the Canadian from his *hacienda* in New Mexico Territory. They were on a two month long hunting trip, not for food or money, but apparently for the sport of it, like English gentlemen out on a big game hunt.

Though he asserted that he knew War Bag, language prevented us from learning how.

Later in the day I climbed back into the wagon to check on the old man and Frenchy. War Bag's head was hot to the touch. His body was racked now and again with uncontrollable tremors.

Frenchy was sitting upright, his head tilted back, staring up at the grey sky. I had to touch his shoulder to get his attention. He felt cold, and his eyes were far away when he looked at me.

"*Bon jour*," he whispered.

"You should cover up better, Frenchy," I admonished, leaning forward to prop the remaining buffalo hides over his shoulders. "You could catch pneumonia." I myself was feeling the cold, having lost my hat. I had to tie my bandanna around my ears to keep the wind from gnawing them away.

Frenchy's hands slipped out from under the hides and patted my back as I tucked him in.

"I won't," he croaked.

"There's a little whiskey left. Sure you don't want some?"

The Frenchman shook his head. His face was clammy from the rain and his wound.

"How about water, then?"

Frenchy's eyes angled up towards the heavens.

"I have all the damned water I need."

The rain died down to a spatter, but the wind was cold and obnoxious. We plodded across frigid, muddy ground, and prayed the wagon would not get stuck.

Near half past five o'clock one of the Mexicans sighted the Indians.

They were just a scattering of black dots behind us, but the sight of them set off a general commotion in all of us. Orfeo seemed strangely calm and assured.

"*No se preocupen, señores.* Where we go they will no follow."

We sighted the north bank of the South Wichita Fork, and there was as welcome a sight as I had yet seen. A camp of white buffalo hunters were crouching under their lean to's like nomads in a grassy desert, watching the rain.

That was probably the only reason the Indians let us see them. If not for the hunters, they surely would have ridden up and killed us all.

Orfeo and Roam hailed the camp. One lone man in a shaggy coat got out from under the cover of the canopy and mounted up. He rode out to meet us.

"What's all this?" he asked, when he was close enough to talk. He was the tallest fellow I had ever seen next to Jack, at least six foot six and solid as the dollar atop a big American horse. He had a Winchester cradled in the sharp crook of one of his long arms, and a big Texas hat with a mustache to match.

"The Tyler outfit outta Colorada," Roam explained. "We was jumped by Cheyenne 'couple miles back. 'Lost our muleskinner and our camp wagon."

The big man looked sideways at Roam. His lip seemed to curl, as though Roam's blackness were a thing to be smelt.

Fuke got down from the wagon and sauntered over, his mouth interceding before he himself had arrived.

"We might've lost more than that if these *Mexicans* hadn't shown up. We got two men *wounded,* includin' our boss."

The big fellow tipped his hat back on his head, peering over Roam's shoulder at Fuke and the rest of our bunch.

Roam's face fell slack, assuming a look of impassivity. His eyelids drooped and he pursed his lips, as if he had forgotten his station in life among our company.

I crawled down from the wagon and came over.

"Pat Garrett," the stranger said to Fuke, ignoring Roam utterly.

"Fuke Latouche," said Fuke, shrugging past Roam and offering up his hand to the tall man. "What *outfit* did you say this was?"

"Skelton Glenn's, outta Griffin," Garrett answered, reaching down from his horse and shaking Fuke's hand. "We're headin' back there in the morning for supplies. How about these Indians?"

Fuke half turned and pointed to the dots on the horizon.

"There they are right *there*," he said.

"I think they've given up," I ventured.

"Well, can't be sure with Indians," Garrett said, his bright blue eyes squinting at the distance warily. "There's a lot of troops out and about, and the fort's not far. Ought to make 'em think twice, anyway. Why not come on down? Skelton'll let you ride with us. We'll get some vittles for your men."

For *Fuke's* men?

"We'd sure be *grateful*," Fuke said.

"Yassuh . . . boss," Roam whispered dryly.

We went the round of introductions (except for Roam, who excused himself to ride to the back of our train and watch the progress of the Indians). Garrett gave us all a good strong grip, ending with Jack, who got down from the driver's seat and came over to exchange gentilities.

The big Missourian had just stuck out his hand when the shot broke out from behind us.

We all jumped and grabbed at our guns, sure it was the Indians. That was when I found that my Volcanic was not in my holster.

We found it in the wagon, clutched in the cold hands of the Frenchman. He would cry out no more.

Fourteenth

We Buried Him in a field of browning sunflowers within sight of Skelton Glenn's encampment.

Jack and Roam dug the grave. It was half-filled with rain water before we managed to lower Frenchy in, bundled in the hide in which he'd died. I do not know that Frenchy was a man deserving of the attentions we denied our more beloved comrade. Maybe his funerary service was in spirit a substitute for Monday's.

The wind wailed sorrowfully in our faces, like a widow with a dead babe at her breast.

Our outfit's only Bible had been left in the coat pocket of its owner on the prairie. Skelton's band had none, so we had to be content with Roam singing the old song he had sung around the campfire so long ago:

> *Steal a-way, steal a-way, steal a-way to Je-sus,*
> *Steal a-way, steal a-way home,*
> *I ain't got long to stay here.*
> *My Lord calls me,*
> *he calls me by the lightning,*
> *The trumpet sounds within-a my soul,*
> *I ain't got long to stay here.*

We didn't know Frenchy's Christian name. Maybe it was just as well, for we had nothing more than a big piece of kindling with which to mark his grave. He was stripped of his guns and gear. I got his rifle, and Jack laid a claim to his pistol.

When our service was over, Roam handed me the Volcanic. I did not want to take it, but I did, replacing it in the gaudy holster. It's weight caused the belt to sag, as the knife which had counterbalanced it was gone. The gun felt strange. More powerful and beckoning. It had killed a man, almost as if it had sensed my doubt as to its worth, and had sought to prove itself to me.

When we came into camp, Fuke, who had not attended the burial, was engaged in talk with our host, Skelton Glenn. Glenn was a chronically serious man with bushy eyebrows and unkempt hair, somewhere near forty.

Besides Garrett, there was an Irish skinner called Briscoe, a bandy legged teamster named Hodges, and a quiet Pole with a wispy mustache the men just called 'Skee.'

The Mexicans pulled War Bag out of the wagon and laid him under Skelton's canopy near the spluttering fire. Orfeo knelt over him, prying at his lips with a wooden spoon of camp stew.

The old man's one eye fluttered like a caged bird beneath the lid.

"Does anybody here know Spanish?" I asked, as Jack and I came over and sat down.

"I know a little," Garrett said.

"Can you ask him how he's doing?"

Garrett did, and Orfeo answered that War Bag was a strong old bastard. So long as the blood in his veins did not go black, he would live. Through Garrett, I got that Orfeo had known War Bag many years ago in Arizona. Apparently one of the old man's nephews had been taken captive by the Apaches, and War Bag had been instrumental in the boy's recovery.

A bit of stew drizzled down War Bag's chin, and I moved to catch it with my hand.

Orfeo smiled at me, and his eyes fell to the gun on my hip.

"*Tu eres zurdo, ha?*" he said, his eyes smiling, as he pointed to his own hand.

"What's he saying?" I asked Garrett.

"He wants to know if you're left handed."

I nodded.

"That's right. *Si.*"

"*Ay, muchacho!*" Orfeo said, slapping his hand on his knee and shaking his head, laughing. "*Los zurdos no se van al Cielo!*"

I smiled as one uninitiated to a joke smiles. I looked at Garrett.

The big man shrugged.

"He says left handers don't go to heaven."

"I can't figure how a pack of Dog Soldiers could've slipped through the Army," said Glenn.

"Where was it you got hit again?" Garret asked.

"A couple of miles south of the *Pease*," Fuke said. "They would've *bushwhacked* us, but one of our men, ah... *surprised them.*" He looked at me with a knowing smile.

It was strange how suddenly Fuke had been elected to speak for our outfit in War Bag's absence. He had not been so outspoken with the Mexicans. Yet now that we were among white hunters, Roam's importance in our group had suddenly diminished. Roam bowed out without a word. I looked across the camp and saw him helping with the lean-to the Mexicans were erecting. Glenn's men ignored Roam as well, and Orfeo was only tolerated in his capacity as an attendant to War Bag. No one in Glenn's group had offered either of them food.

I got up and went over to pitch in with the raising of the shelter.

The Mexicans looked a little queerly at me when I offered to help, but in short time we had a length of tarp to keep the rain off our heads. Roam went and got one of the bundles of dry kindling he kept wrapped in canvas.

"They've got stew over there," I said to Roam, meaning Glenn's camp.

"I know," Roam said, spreading out the sticks. The dark stain of Monday's blood was on his tunic.

One of the Mexicans came forward with a bit of flint and steel and got a fire going. From their satchels and saddlebags the men all produced some kind of food—breads and rolls, washa, beans, jerky, tortillas, and spices. Everything was passed around as it was needed. I knelt under the canopy next to Roam in silence, for I had nothing to offer.

"Why don't you go have some?" Roam asked, pouring water from his canteen into a pot for boiling.

The rain pattered on the material just above my head, and the smoke from the fire stung my eyes.

I got up and went back to Glenn's fire, passing Orfeo on the way. He was going to join his men and Roam. I started to thank him as we passed, but he spared me no glance.

"We're about two days from Griffin," Glenn was saying, as I sat back down. "With all them soldiers out lookin' for them German sisters, I warrant them Injuns'll give up their chase."

"I'm surprised they got through the patrols at all," Hodges said.

"What German sisters?" I asked as I sat down.

"You boys have been out of touch," Glenn remarked. "'Bout a month ago some scouts outta Ft. Wallace come across a family of five and a burnt wagon. They found the family Bible, only there was nine names in it. Injuns run off with the four little girls."

"Not a one of 'em over seventeen," Garrett said, shaking his head and sipping a cup of fresh brewed chicory. "But them soldiers were already in the field when it happened. They're as useless as a new suit on a hog."

"General Miles has got three columns out there lookin' for 'em," Glenn said. They figure it was Cheyennes. Medicine Water's band."

"Unless they been traded," Briscoe chimed in.

"At any rate, it's hard luck bein' an Injun right now," Glenn said.

"I hope they gut that son of a bitch if they catch him," Hodges growled.

"Oh, they will," Glenn said, but if he meant to assure that they would catch this Medicine Water, or that his punishment was so ordained, he was not plain.

"Well, them *Cheyennes* that jumped us," Fuke said. "There must've been a dozen or more. We accounted for more than *half* their number."

How he knew this, I had no idea. I had not got a sure count on their number, and had only counted four dead.

"Cheyennes don't range this far," Garrett said.

"Well our scout says they were Cheyennes. Their chief had a mark on his face," I said. "Like a powder burn."

Glenn looked over at me, interested.

"No foolin'?"

"I saw him myself," I admitted. "He's the one that jumped War Bag."

"That might be ol' Black Face," Glenn said. "I tell you 'bout him, Patsy?"

Garrett shrugged and drank his chicory.

"Mean son of a bitch. He was one of 'em that holed them fellers up on that sandbar in Colorado...where Roman Nose got rubbed out."

"Beecher's Island?" I asked, a little excited.

"That was it," Glenn nodded, stroking one of his baroque eyebrows and ruminating. "They say he come outta Sand Creek. He rode with Tall Bull and Bull Bear, some of them other bad 'uns. He was at the Washita too, against Custer. I wouldn't be surprised if he didn't have something to do with the killin' of them surveyors at Lone Tree, 'er them German sisters either."

Garrett looked at me and rolled his eyes, as if he'd heard Glenn's Indian theories a hundred times or more.

"Hell," said Glenn. "Black Face. You boys are lucky. That murderin' savage has not got a drop of mercy in him."

"They had a Negro with them," I said. "He looked like a soldier."

"Don't surprise me," Hodges said, dishing out stew from the big pot on the fire. "Way I hear it told, them nigger soldiers is quittin' the Army and joinin' the Indians faster than they can get 'em trained."

"I always said giving *guns* to niggers was bad *policy*," Fuke said, taking a bowl from Hodges.

"How 'bout that one you got with you?" Garrett said, having drained his chicory. "He a deserter?"

"*Probably,*" Fuke said.

The men around the fire looked at me, and I felt my face grow hot. Fuke supped his stew and smiled at me in a patronizing manner.

I shook my head. I looked back across the camp as the men's talk turned to other things. Roam and the Mexicans were supping. I realized they had raised the tarp for themselves. Their regulation to another part of camp was a thing unspoken, but clearly understood. This wasn't Cutter Sharpes,' and War Bag was not able to vouch for Roam.

As Fuke and his new comrades talked and laughed, I caught Jack looking at me over an empty bowl of stew.

"Jack," I said. "Do you think Roam is a deserter?"

"Naw," Jack said right away. "Ah reckin not."

I bedded down. I couldn't dismiss the notion that Fuke would not have spoken so about Roam if either he or War Bag had been around. I fell asleep watching the Mexican fire. Roam was alone with strangers. I didn't eat that night.

Fifteenth

The Gradual Segregation of Roam and the Mexicans was absolute by the time we reached Griffin.

Fuke struck up a friendship with the tall Pat Garret, who turned out to be from Alabama. They spent the next couple of days trading sizable whoppers and pocket money over cards, until they deduced it would be mutually beneficial to forge a partnership and bilk the Irishman and the Pole out of their money instead (Jack too, until I warned the big man away from their treachery, earning a disapproving silence from them both for the duration of our trek).

I spent most of my time with Jack and War Bag, and of the two, I cannot decide whose conversation was less interesting. War Bag's fever broke thanks to Orfeo's constant care, but he remained restless and addle-brained.

As to Orfeo, our exchanges were infrequent due to the language difference. I managed to talk Garret into translating a few more times before his annoyance at my interest in what the Mexican had to say became intolerable for both of us. I learned little that was new from him. War Bag's progress, the names of some of his men.

Trigueño was the nephew that War Bag had saved as a

boy from the Apaches. He was an angular youth of not more than seventeen, in a faded red serape. He had a shapeless heap of dark hair and small, walnut eyes. His skin was so brown and his face so wide, I would have thought him an Indian. He spent most of his time alone, watching the campfire or the land, sometimes helping his uncle tend to War Bag. At such times, he fixed his narrow eyes intently upon our convalescing chief, and his thoughts were hidden behind an unreadable mask of detached intensity. I found him unnerving.

Roam continued to share time with the *ciboleros*, and I frequently found myself standing around the edge of their group like a shy schoolboy. Roam's understanding of their language grew, and soon they were swapping jokes and laughing.

We came across a weird expanse of dead mesquite trees about a day and a half outside of town. They were bleached white in the sun and littering the landscape like the long discarded bones of some unfathomable creature that had long ago shaken off its mortal coil. They rendered the land almost impassable, and were disconcerting in their sheer number and in the air of supernatural dread they lent to the barren land. The earth was very dry, and nothing lived. Our horses resisted our efforts to cut through, as the claws of the brittle bushes raked at their flanks (and our legs). We had to circumvent them, and it added time to our journey.

The town of Griffin sprawled in the valley of the Clear Fork, beneath the rise of Goverment Hill. The fort proper was little more than a few low adobe, log, and store buildings. It occurred to me that the Army liked to build their forts on hills, and the towns that sprouted up around them were like feudal villages lying under the protection of a castle. Hugging the eastern bluff of Government Hill was a village of huts and dingy, conical tepees whose points could be seen jutting up here and there between the picket buildings of the town. This was the home of the local tribe of Tonkawa Indians.

The Flat, as the town was referred to by the soldiers and civilians alike, was a cluster of buildings both rude and refined, picket and adobe, lying on both sides of the Brazos crossing. It was a community of diversity if not (at least not yet) proper respectability.

There was an industrious tannery, and the yards surrounding it were covered with stacks of hides fifteen feet high. This lent a smell to the town that could only be easily detected by the outsider. It was the smell of dead buffalo, and we were too used to it to notice.

The only clearly defined street was Griffin Avenue. It was rutted with the passing of hundreds of laden freight wagons and horses, bearing loads of hides and cottonwood lumber in from the woods and plains and out to the railhead at Denison. The avenue continued east of town and branched off northeast and south, before dwindling into the tree-lined west bank of the Clear Fork of the Brazos. Tall wild sunflowers lined the road in, but they were brown as the ones around Frenchy's grave.

Griffin was a town on the brink of a boom. The Western Trail had just reached it, leaving a few bewildered young cowboys in its wake, and it was also seeing the immigration of the buffalo runners down from Kansas. It was a relatively lawless, rowdy town, glutted with adventurers and soldiers. By lawless I do not mean to conjure up romantic images of gun duels in the streets and masked bandits besieging the savings and loan (there was no savings and loan). Nor do I infer that this was some sort of Sodom on the plains, as I have heard it called by Temperance ladies in years since. I only mean that the law generally had other things to do than parade around town, and the military mainly concerned themselves with their own. Henceforth the conduct of the locals was generally less than reserved if not at times downright bawdy.

Before we had traversed the length of it that first day, we witnessed a drunken altercation between a runner and a cowpuncher from one of the local ranches, an impromptu horse race down the main avenue, and the ousting of a young soldier from an establishment of questionable morals.

The soldier was dragged out by his boot heels by a beefy, bald headed man who deposited him in the muck of the street right at our feet and stalked back inside without another word.

As we watched the infantryman stir in the cold mud, a woman's voice rang out from an open window, and we all craned our necks to see a shocking sight.

Leaning on the sill with one hand poised on the faded shutter there was a pink skinned, dark haired woman wrapped only in a multi-colored quilt, her breath puffing out in the cold air. One floppy breast drooped out from the folds of her covering.

"Come on back and see me when your payroll's in, boy—and not a day sooner."

I averted my eyes instinctively, taking a sudden interest in the floor of the wagon. I had never seen a lady's bare bosom before in my life, outside of some grainy daguerreotypes a leering school chum had once procured from somewhere.

But then, I suppose it must be contended that a woman who exposed herself to strangers so nonchalantly could not rightly be considered a lady. At least not any as I had known them.

Fuke, it seemed, for all his high talk of God, culture, and East Baton Rogue, held no truck with such proclivities.

"Hey there, darlin!" he shouted up to the woman in the window. "D'you restrict your attentions to blue chained dogs, or has a *free man* got an equal opportunity?"

I felt my ears and cheeks flush, despite the cold. The infantryman propped himself up on his elbows and began to vomit in the street.

"Well, most of you buff runners ain't heard of a new fangled invention we got called soap," she called down.

The men cackled appreciatively.

"Why, a seasoned man has got a spicier taste, is all," Garrett called.

"Just like Polish sausage eh, Skee?" Hodges chortled, slapping the Pole on the shoulder good-naturedly. The Pole was surly, however, as he had lost most of his money to Garrett and Fuke, and he cast his face in a scowl and shrugged his partner's hand off.

"You just be sure and scrub yore sausages 'fore you come see me," said the woman.

All the men let out hair-raising whoops and calls, and the Mexicans trilled like yipping coyotes. I felt an overwhelming desire to hunker down in the back of the wagon and nurse my blazing ears back to their pale norm.

Fuke must have noticed, for he immediately set out to embarrass me further.

"How about it, *Juniper?*" he said, turning to me. "Maybe we'll get her to wash behind your *ears* for you—get you to forget that little peach *Miss Penrose.*"

In truth, I had not thought of Miss Penrose in some time, but the mention of her name in the vicinity of that bawdy house and that woman seemed like an unforgivable offense. I felt inclined to change the subject.

"I could use a new hat," I said. I had come down with a cold, and had been quite miserable for the past day or so.

"Aw, don't worry, *Juniper*. The whores aren't *particular.*"

"Forget the whores. I'm freezing," I reiterated.

The soldier heaved again in the street and I wrinkled my nose.

"Go to Conrad's," said Glenn. "He's the sutler at the fort. His prices are fair."

*   *   *

Judging by the amount of bustle on Government Hill, the talk of military mobilization on the Plains was more than idle rumor. There were bands of runners and scouts and marching troopers going in and out with regularity, and there was a real air of purpose in the place such as I had not felt back at Lyon.

Frank Eberson Conrad's sutlery was just as crowded. No less than three clerks were at hand. Goods were stacked nearly to the ceiling in some places. Where there were no shelves, there were piles. The corners of the store were stacked with blankets and saddles. One long table was filled with ammunition of all kinds. Another held a cornucopia of spice tins. One niche contained nothing but footwear; boxes of boots and shoes for men and women, with little drawings of the contents on the outside. There were rigs and belts for men and beast. There were kits and saddle boxes. Bolts of fabric and coils of rope. Bowie knives and bullet molds. Slickers and handkerchiefs. Stacks of shirts and heaps of pants. Baskets of bung starters and bushels of barber strops. Scabbards and cases for every conceivable item, from rifles to pocket watches. It was a far cry from Matthew McKintry's little store.

Conrad even catered some to sight seers, although I couldn't imagine that there were many who stopped through. There were little baskets of "Real Indian Scalps," and a pyramid of brown bottles labeled, "Blue Mass Pills" on the counter.

Two of the clerks bustled back and forth with armfuls of goods. There was one man whose sole job was to handle money. He remained stationary at the counter, tending the affairs of the customers who traded him dollars and army script for the right to walk out of the shop with all sorts of things. There was a curtained back room, and yet another man hustled back and forth from here with change from an iron safe that could be seen through a part in the curtain. If any of these men were Frank Conrad, I did not know.

From a stack of hats tall as me, I selected a cream colored, wide-brimmed affair with a stern brown band.

"That nearly cleans me out," I said, stepping away from the counter and fitting the stiff new hat on my wet head.

"We'll get the old man a room and a *sawbones*, and we'll divvy up the cash," Fuke was saying, as we both came out into yard where Jack, Roam, and the others were waiting.

"I don't know 'bout that," Roam said.

Garrett and Glenn looked sideways at Roam, then regarded Fuke questioningly.

Fuke bristled under their expectant gaze. What they expected, I could only surmise.

"Who made you boss of this outfit, *boy*?" Fuke muttered.

Roam's jaw tightened.

The dark eyes of Orfeo and the other Mexicans flitted back and forth from Fuke to Roam, jerking the situation tighter. They spoke this language, even if they didn't speak English.

"War Bag's boss. I figure we don't divvy up till he say so," Roam said, his voice even, but just barely.

"Ah reckin that's the truth of it," Jack said, watching Fuke. He had chosen a side, and now it was my turn.

"It doesn't seem right for any us to split up the money," I said. "War Bag said he'd do it at the end of the season."

"You need him to tell you this season's over, *Juniper?*" Fuke said, not taking his eyes off Roam.

"We all goin' get what's comin' to us," Roam said. "If the man die..."

Fuke took a step toward Crawfish, where the old man's saddlebags were tied.

"If he dies, *you're* going to oversee the split?"

Roam started to speak, but seemed to think better of it.

"Let's worry about all that later," I interrupted. "We can split it together, if it comes to that. Right now, I think we really should wait."

Fuke shrugged. He finally looked at me, as if betrayed.

"You'd spend your whole *life* waiting, wouldn't you, *Juniper?*"

"Hey, where are we gonna stay?" I asked, ignoring Fuke's hateful look.

"Bison Inn's got the cheapest rates," said Glenn. "Ain't much more than a tent, but you could get your own partition by the month or by the night."

"Fuss, we goin' head to the livery," Roam said. "I'll sell off that mule and get stables for the hosses."

"You do that," Fuke said, turning from us.

"Whar ye goin,' Fuke?" Jack called.

Fuke made no answer. He shrugged past a pair of filthy Tonk wolfers and was gone.

"Aw, he jest goin' t'get mean-drunk," Roam said. "Let him."

"I believe we'll join him," Glenn said. He tipped his hat to Jack and me, but again exhibited that aloofness with Roam. "Be seein' you."

Garrett, Glenn, and the rest of their outfit gave heel or rein to their animals and trotted off after Fuke.

Orfeo cleared his throat.

"*Hierba mala nunca muere,*" he said, shaking his head. He reached into his tunic and produced a handful of coins. He thrust them at Roam. "*Mira!* Take this. *Para su jefe.*"

Roam made a motion to refuse. Truly, we all should have been paying Orfeo and his men for the protection they had given us.

Orfeo took hold of Roam's hands and put the money in.

"*Por favor.* For *Senor Tyler.*"

"Awright," Roam nodded. "Awright. *Gracias.*"

I was taken by the Mexican's show of generosity, but more so, I was surprised Roam took the money. Hadn't we enough of a stake to share War Bag's expenses?

"*En cuando el se levante,*" he went on, straightening in his saddle and removing his grand sombrero, "*encargate de darle los complimentos de Orfeo Largueza.*"

"*Yo voy,*" said Roam, smiling. "I will."

Orfeo grinned and replaced his hat. He nodded to Roam, then saluted us in parting. "*Adios, señores. Vayan con Dios.*"

They spurred their horses and went off across the yard, bouncing colorfully like visiting spirits in the dull browns and greys prevalent in the fort. The wild Trigueño was the last to pass. He stopped beside the wagon and cast a lingering look down at the old man, before taking off after his fellows.

"Them Mexes ain't all bad, are they?" Jack said.

"Naw, not even half," said Roam, putting the coins into his pocket.

We went back down into the Flats and to the livery, where we spent some time haggling over the going price for Michael the Mule. Roam managed a fair bargain and accommodations for our animals as well. As we left the mule behind, I could not dismiss the feeling that I was in a way leaving Monday as well.

I expressed this aloud, and Roam shook his head.

"When my time come, I pray don't nobody need a jackass to remember me by."

We checked into the Bison Inn, which was more fit for bison than for bison hunters. It was basically a frame and canvas shelter divided and subdivided by muslin flaps tacked floor to ceiling. Each little room was lit by a tin oil lamp and consisted of a dingy pale for refuse and a spot on the sawdust floor to spread a pallet (or a potato sack could be rented for fifteen cents a night). There was not much privacy and the floor was jumping with bed bugs. When I inquired about bathing facilities, I was directed to one of the whorehouses which had one of the only two public bath tubs in town (the other belonged to the slightly more upscale but unimaginatively named Bison Hotel).

Adjoining the living quarters was an eatery with three or four long tables, two kegs of nickel beer, and a lunch counter of bread and buffalo meat (which looked to be a little ancient and heavily salted).

Hard stares were directed at Roam upon our arrival, and he murmured something about waiting outside. We went in and secured rooms for ourselves.

We got War Bag a premium accommodation (premium because it was located nearer to the pot bellied stove next to the center pole which served to warm the entire place) for a dollar fifty a night, and then settled in to our own cubicles for a dollar each (fifty cents for the accommodations, we were told, and fifty for the use of the lunch counter).

Through the partition, I heard another new arrival remark,

"There ain't enough room t'cuss a cat in here without gettin' a mouthful of hair."

"This whole place has probably got the black plague," I muttered to Jack. "It should be burned."

"Aw, it ain't so bad," said Jack.

"Not bad?" the shadow of the man next door exclaimed. A whiskered face poked through the corner of the flap that separated his abode from mine. "Why, I've bedded down in prison cells with vomitous drunks for companions that were cleaner than this. I shit you not."

The face withdrew and was replaced by an out thrust hand.

"Name's Bill Russel. Everybody calls me Windy."

I shook his hand, and we went the round of introductions. He was a talkative skinner from St. Louis, and his partner was a mournful looking fellow named Jim White.

"Don't take your boots off tonight, boys," Jim warned us gravely. "Lord knows what'll be there to greet your feet in the mornin.'"

After Jack and I were squared away, we went out and found Roam waiting patiently with the wagon and War Bag.

"Why don't you an' Jack go try t'find Fuke," Roam suggested. "I'll see 'bout a doctor for the ole man, and get the wagon and the oxen stowed in the yard."

That meant venturing out into town and rubbing shoulders. Yet I was not very keen on bedding down among vermin, tired as I was. I suppose walking among any sort of human kind was preferable.

"Reckin' a drank would sit well with me," Jack said.

"What about you, Roam?" I asked. "Where are you going to stay?"

"I think I might have me a place," he said.

"Well, alright," I told Roam. "We'll meet you later, then?"

"Later," he agreed, and urged the wagon bearing the old man forward.

I ducked back inside and asked Jim White if he would like to accompany us. He answered in the negative. He was tired, and would be glad to get a full night's sleep without fear of Indians.

"How about you, Bill?" I said to the dirty flap of canvas beside me.

The only answer from Windy Bill was a long and contented snore.

Sixteenth

Griffin had two saloons at that time. Jack and I searched them both for Fuke.

The first stop was Shaugnassey's, a reserved place full of locals and conducive to easy talk. Between glasses of dark beer we traded a few words.

"Jack, why'd you come to Texas?" For our months of travel, I really knew very little about him.

"Follered m'brother Tom."

"I didn't know you had a brother, Jack. What did he do? Was he a tanner as well?"

"Naw. Tom never took to workin' in the tannery with me an' paw. He come out here on account o' most o' his friends was lightin' out fer Texas after the War. They was Black Flaggers, y'see."

"Border Ruffians," they'd called them in the Northern papers. Looters and killers of women and children. Not part of the regular Confederate Army.

"You were a Black Flagger too?"

"Naw. My kin didn't have much to do with the War, but Tom's town friends did. They could nigh talk him into a holler log fulla bees. When he come back after the War, he whar a wanted man and most alla his town friends was dead. Thar was Fed'rals all over the low country lookin' fer them that'd rode with Quantrill and Anderson. Tom opted to go to Texas, and Paw asked me to go on and watch over him, me bein' the eldest."

Jack sipped his beer, suddenly morose.

"So you left home for your brother's sake? But what happened . . . ."

Jack just looked at me with his sorrow-filled eyes, and I knew Tom McDade would never go back home.

"I gets to thinkin' o' Brown Mountain," Jack sighed. "An' sometimes I gets t'pinin' an' feelin' way down past low. Paw's tan yard, the busthead still . . . that ol'black bar . . . alla my 'lil cousins what used to hang on me at gatherin's."

"Why don't you go home, Jack?"

"Ain't no goin' home for me, Stretch," Jack mumbled, "without Tom."

When we swaggered out into the waning dusk, Jack suggested we try The Bee Hive Saloon up the street, so we made for its jaunty swinging sign. All the while I swore to myself that after a quick look, if we saw no sign of Fuke, we would move on. But when we ducked into The Bee Hive and saw the hurdy gurdy girls dancing in their frilly outfits, my resolve was washed away like so much driftwood.

Guiltily, Jack and I slowly filed behind the other wretches who were crowding and craning their necks to see inside that den of sin.

> *In this Hive, we're all alive;*
> *Good whiskey makes us funny.*
> *If you are dry, step in and try*
> *The flavor of our honey.*

The sign swung mockingly over our heads from a squeaky chain. The rhyme was inscribed over a crude yellow-painted rendering of a beehive.

It was just one floor, but widespread. There was a little stage in the back with a gaggle of tables facing it. On stage were four girls, shockingly exposed in raucously designed stripey yellow and black unmentionables which flaunted their womanly aspect in brazen disregard for propriety.

The tables were mostly infested with goggle-eyed men, drunk and hollering blearily up at the performers. Those whose attentions were not riveted to the entertainment played at cards. The brightly colored chips were the only thing to rival the stage among the dull, dirty hues of the audience.

In one corner, to the right of the stage, three musicians picked at a twiddling rendition of 'Sally In The Garden' with banjo, fiddle, and washboard. The girls kicked and pranced. Weaving in and out amidst the sea of beer-stained tables were about six more uniformly dressed (or underdressed) girls dancing closely and in some cases in a most unseemly manner with lusty male patrons.

A long bar was manned by two officious looking distillers who staunchly held back a phalanx of men hammering the flats of their hands on the counter top for service. Drink orders spouted from the clumsy lips of men who had in some cases already made a few trips to the bar.

Jack and I stepped in and began scanning the overcrowded room for Fuke.

We were making our way through the crowd when the music stopped most abruptly and the ladies on the plank stage moved in an orderly line to a row of stools. While I was watching them float across the room, the fiddler took his chin off his instrument and called out in a voice that was sing-song from long hours of repetition.

"Courtesies for the ladies please, gentlemen, courtesies for the ladies."

The girls who had been shuffling about on the dance floor led their stumbling partners by the hand to the bar in one body. Those patrons already there parted as the girls approached. The bartenders had begun setting up a row of glasses as soon as the music stopped. By the time the girls and their partners lined up, they had filled them all from the same bottle.

Jack and I were accosted by two women whose attire made me blush.

The girl who gripped my arm was a willowy, green eyed thing with high pile of wiry blonde hair and a dusting of freckles over her bare shoulders and neckline. She was a bit shorter than me, and as I looked down to see the source of the light touch at my elbow, I found my eyes plunging down the spotted crevice between her swelling bosoms.

"Hey there good lookin'! How about a dance?" she said brightly.

At the same instant I heard a keening little voice say to Jack,

"How's about you an' me go an' dust the floor awhile, Tree Top?"

Mine was not much older than me. She was thin with an ostrich-like neck. Her teeth were crooked, and there was a fine tracing of hair on her upper lip that shimmered in the lamp light. I found myself nodding a dumb assent after dragging my eyes up from her ample cleavage to stare stupidly into those shining green eyes.

Before I could do more, her hand slid down my arm and encircled my wrist. She led me right through the maze of men and tables towards the dance floor. I imagine that Jack was given the same treatment, though I did not notice. My attention was fixed on her form as she led me. She wore fanciful yellow cotton bloomers which did little to conceal the swell of her womanly hips. This garment billowed out and then cinched tightly around her limbs just above the knees, exposing a tantalizing bit of white flesh between the cuffs and the rims of her black stockings. Her long limbs narrowed as they dove into a pair of clunky shoes, and I noticed a white slash of running fabric on her left calf. My throat was dry and my clothes too tight, the pit of my stomach warm and swimmy. She looked over one bare shoulder and smiled a pretty, close-lipped smile.

As we made it to the center of the room, the band struck up a slow waltz. The girl placed my trembling hand on her hip and slid her grasp from my elbow to my sweaty palm, upraising it and bending my elbow properly.

"It's like this, see?" she said.

"I know," I mumbled, and began to lead. My mother had been sure I got plenty of instruction in dancing as a child, and I had waltzed numerous times. I do not know the wordless, inexpert tune the band played, but I found the tempo nonetheless, and guided the girl about the floor. It was difficult, as the shoes she had on were large and awkward. She stumbled a few times before stopping. She balanced herself on my shoulder and reaching down, hiked up one leg and popped the offending footwear off one at a time. But after she had accomplished this (rendering her about four inches shorter), we only had time for four more turns before the song abruptly ended, and the fiddler called out again,

"Courtesies, gentlemen. Courtesies for the ladies."

All around us couples began rushing to the bar as before. My companion reached down and picked up her shoes. She smiled at me prettily again, the pinkish sheen of her gums just visible between her lips, her cheeks puffing up charmingly.

"That was real nice, mister. You shore know how to dance."

"Thanks," I muttered, forcibly keeping my eyes locked into hers. It would have been ungentlemanly to let them rove while she was speaking.

"Well, come on. You better buy me a courtesy."

She led me to the bar. I was sandwiched between her and another girl, the smell of lilac water and ladies' powder heavy in my nostrils

"What would you like?" I asked.

"Oh you gotta get me one of them lady's specials. Don't worry, he'll give it to ya."

The professor came over, and before I'd even asked, he had a shot glass half-filled with a rust colored liquid before my dancing partner. She drank it down in a gulp, without a grimace or pause.

"Anything for you?" the barkeep said to me.

I turned to the girl at my side.

"Are we going to dance again?" I asked.

"Shore!" she smiled happily.

I returned to the waiting bartender.

"A shot of bourbon, please."

The bartender produced a labeled bottle of bourbon quicker than I thought possible, and tipped it into another shot glass, cutting it short from the rim by about an inch and not spilling a drop.

I reached for the tumbler, sparing the bartender a disapproving glance which did not seem to affect him in the least.

"How much?" I asked, reaching down into my ever-lightening pocket.

"A dollar and fifty cents," he said.

My hand jarred slightly as it closed around my meager funds.

"Dollar and fifty!" I exclaimed. "What...?"

My partner began tugging me away from the bar.

"Come on, mister. The music's gonna start."

I plunked the money down. My drink sloshed in my hand and I'd barely thrown it down before she had me out on the dance floor again. As I was led off, I glanced back at the bar. The professor was wiping it down with a towel. The glasses were already gone. As I blinked the space filled with three or more pressing men hollering for libations.

"What the heck was in your drink?" I asked my companion as we returned to familiar territory.

She drew me in closer than before, and I felt her body press against mine. She had to stand on tip toes to speak into my ear.

"It's just sugar water."

She pulled my arm all around her waist and hugged me close.

"Dance with me some more like you done."

The band played a melodic schottische, and we twirled about. The bourbon and the Shaungnassey beers went to war somewhere far down in my stomach. I managed not to step on her toes, which I felt was a great achievement considering.

Meanwhile the closeness and the scent of her was doubly confounding to my impaired senses. I felt the bare flesh of her shoulder blades beneath my fingertips, and the swell of her half-exposed breasts against my chest. She leaned her head against my shoulder, and I smelled the fragrant powder in her straw-colored hair. I was infatuated, but deep in the back of my mind, a nagging thought was eating away at my pleasure. I could not afford another dance. I was busted.

"You shore dance grand," I heard her muffled voice say against the folds of my coat. I was aware that I must be radiating an unfavorable bodily odor, but she didn't seem to care. Maybe the heavy scent of her own perfumes and powders was cancelling out my own, or she was so used to the stench of the patrons of the Bee Hive that she had ceased to take notice.

I smiled and let my chin rest lightly in her curling hair.
"What's your name?" I murmured.
"Cathy," she said. "What's your's?"
"Stretch," I said.
She giggled. "You don't dance like no Stretch. You dance like a Walter or a...or a....Marmaduke."
I snorted.
"Marmaduke?"
She giggled again, and it was beautiful.
"Yeah. Like a fancy gentleman with a twenty dollar name. Come on, what's your real name?"
I told her. First, middle, last, and the number on the end.
"Yeah. That fits."
The band began to wind down.
"Aw," I said.
"What?"
"Cathy. I don't have any more money."
She put her head against my chest again.
"It's alright. Just dance with me a little longer."
Suddenly I felt like crying.
The tune ceased, and the fiddler called out mercilessly.
"Courtesies, gentlemen. Courtesies."
Cathy and I parted, and she smiled. She braced one hand on my shoulder again, and slid back into her ugly shoes.
"Ain't used to such a nice dancer," she said.
"Maybe I can borrow some money from my partner. Will you wait for me?"
She smiled, and straightened as she fixed her shoes on.
I rushed off like a kid to find Jack. I bumped right into a man headed for the dance floor, and he stumbled against a poker table, upsetting a few chips.

"Goddammit!" one of the poker players shouted, standing up so abruptly in his anger that he overturned his chair.

The man I'd upset shot a look at the gambler, and the card player's face seemed to visibly lighten from red to a pasty white.

The little gambler raised his hands up.

"Awright, Jinglebob. Didn't see it was you."

"It's awroight, is it?" growled the man I had jostled. He had a musical lilt to his speech that I could not place. It was like the bleating cockney of Cutter Sharpes, but smoother.

"Shore," said the gambler, eager to return to his game. "It weren't your fault."

"No. It weren't," the foreigner growled, turning to face me.

Jinglebob was the ugliest man I had ever seen. He was an albino, and the pinkish blemishes in his skin stood out starkly on his yeast white face. He was square-jawed, with big blocky horse teeth, even and straight beyond belief, as though they were cut from limestone. His nose was piggish, and his pale blonde eyebrows were like thick awnings of faded wool over his glaring pinprick yellow eyes. But the truly hideous feature about him was his right ear. The lobe and part of the lower half of it had been partially severed, and hung only by a scrap of scarred flesh. When he spoke the torn lobe dangled and bounced sickly on its thin tether of flesh.

He was big and broad shouldered, though a little short. His hair was scraggly and hung under a city-style derby. Two strips of furry sideburns framed his intemperate face down to the chin. He wore a big horse pistol, and suspenders to help support its weight. The sleeves of his red flannel shirt were rolled back to reveal tenpin arms clean of hair and thick as mutton hams. He leered at me, and took one too many steps closer, until we were eye level. I had to force myself not to stare at that quivering, ruined ear.

"It's *your* fault, mate," he hissed at me. I felt the hairs prickle on the back of my neck. His breath was stale. "Where the hell are you goin' to in sawch a bluddy hurry?"

His accent was like listening to a man talk around a bitter lemon half.

"Sorry . . . ," I stuttered.

Jinglebob looked me up and down. His eyebrows were up like bent bows, as if open to any fight I might put up. I offered none.

"Heah now. Accidents'll heppen," he said. He slapped me hard on the arm, and it stung. My face broke into a shaky smile. He laughed, and I saw a great gap in the top row of his teeth. "Boigh me a drank and we'll forget awl about et."

My smile faltered.

"I don't have any money," I said.

"Sure you do! Check yeh pawkets—a fella wouldn't com into a place loik this without a bit o'dosh."

I shook my head, and turned my pockets inside out to show him.

"Look," I said dumbly. "I'm sorry, I just don't have anything."

Jinglebob stared at me, and sighed heavily.

"I was going to see if a friend of mine would lend me some money . . . ," I said quickly.

"Oh yeah?" he said. He turned to the two men who were with him.

They were both grinning ear to ear, enjoying the bullying. One was a tall, good looking cattleman with a thick moustache, and the other, an unkempt fellow with robin's egg eyes and an unshaven neck. They were both armed.

I found myself pulling the bottom of my coat closer to hide the pistol on my hip.

"Well, whoigh don't yeh go and foind him and we'll wait heah. Yeh'll boigh me mates a round too, roight?"

"Sure, sure," I said, nodding eagerly and backing away. "I think my friend's outside though. I'll just go to him and be right back."

"She's roight, she's roight," said Jinglebob, his ear twisting like an obscene Christmas ornament. "Go on. But be sure and come beck, now."

I nodded dumbly, backing away the whole time, until I was stopped in my progress by a man seated at another table, whose hat I knocked off.

I hastily picked it up, not wanting to offer further offence to anyone else. I heard Jinglebob and his cronies laugh derisively as I returned the hat to the card player and made my way to the front door.

Out in the street night was upon Griffin. A rowdy tune skirled out from behind me as the band began anew. The shops were closing down and the saloons were lighting up. Soldiers down from the fort swayed on the walk, laughing to each other and heading for perdition. I did not see Jack. I stepped cautiously back into the doorway and looked over the growing crowd, afraid I'd catch Jinglebob waiting for me.

In sickly heartbreak, I saw the contemptible mangle-eared ruffian out on the dance floor with my Cathy. The two of them moved lasciviously against each other while his friends sat at a nearby table and clapped their hands in time to the ribald music.

I leaned against the wall outside, feeling low and sickly from the cheap bourbon and beer. An unreasoning hatred began to burn in me for the bullying foreigner who had made me look and feel so foolish, and who now was enjoying the attentions of my hurdy gurdy gal.

I touched the Volcanic at my side. My thoughts fell on the Indian I'd killed, and the indelible picture of Monday Loman as he breathed his last painful breaths, and on Frenchy's body curled among the hides with a hole burned in his temple. I felt sick, and the bad liquor in my moving belly seemed to run down into my legs. I reeled and fell against the hitching post. My head was pounding and my bowels felt cramped.

I turned away and saw Fuke and Pat Garrett entering a building up the street. Swallowing a thick lump in my throat, I meandered over.

A red paper lantern hung outside, casting a hellish glow on the doorway and the muddy street in front.

A gorilla like man reared up from a stool in the shadows of the wall and blocked my way. It was the big bald man we had seen oust the sickly soldier, and I realized this was the same place we had seen on the way into town. The bald man's face was pockmarked, and his eyes black in the night.

"You got money, kid?"

"I'm just looking for a friend..."

"Friends cost money here."

"I mean the fellow who just went in . . . ."

"You can wait for him out here, if you want."

He put one big widespread hand on my chest, and I glanced down at it.

"You don't understand . . . ."

He shoved me back hard, and I slipped off the short boardwalk and landed on my hind end in the filthy road. Somewhere across the street I heard a woman's strident cackle.

"Get outta here, boy," he growled.

I was infuriated. Without thought I grabbed at the gun at my side.

Just as I did so, the big man turned his head towards the door. It had opened a crack, and a woman's voice, harsh as whiskey in a cracked china cup, spoke.

"Somethin' wrong, Barney?"

My eyes widened at the feel of the pistol in my hand. The red light glinted on the barrel. I quickly put it away, my heart pounding in my chest. The nausea welled up in me again.

"Just some kid," Barney said, looking back at me with scorn, oblivious to how close he had come to stopping lead.

The door opened all the way. There stood a woman about twenty years my senior, swaddled in nothing more than a black corset, a pair of lace stockings and a floral print robe which she jerked open before my eyes. Her dark hair fell in unkempt strands around her rippling, fat shoulders. Her red lips were a thin, mirthless slash across a haggard, ruddy face heavily painted with rogue and powder like that of a chipped old doll's. Her eyelids were nearly black with cosmetics, and hung like drawn shades over dull, soulless eyes. I was shocked at the sight of what she so prominently and carelessly displayed between her full hips, and what hung sadly down over the lip of her corset, barely concealed by the lapels of her faded robe. In the angry light she looked like some terrible thing sent to torment the souls of Hell with its loveless form.

"Well there you go, kid," she said to me in that rusty gate voice. "Anything else you gotta pay for."

Barney nodded appreciatively. He looked from her grotesque form to me, his ugly face splitting wider at my agape expression.

Over her shoulder I saw Pat Garrett standing and taking off his hat. There was a lean, brown skinned Mexican girl dressed similar to the woman in the doorway. She was helping Garrett out of his coat. He glanced up as Barney broke into a mocking laugh and the woman shuffled back inside, closing the door.

Barney continued to laugh and shake his head at me as he regained his seat in the shadows by the red door.

He laughed as I picked myself up out of the mud and horse shit. I found my new hat and placed it back on my pounding head. I wanted to kill him. I wanted to kill him so badly I was shaking with it. I glared at him, trying to project some sense of the death I was envisioning for him in my mind, but he kept on laughing. I had a gun. I could have killed him.

Instead I went away with his laughter in my ears, swirling like a ghost wind through my body, under the arcs of my soles like static lightning. My hands trembled at my sides, and my eyes swelled with tears.

I saw Jack, standing amidst a crowd of people across the Avenue and back towards the Bee Hive. I shoved roughly through the crowd, as though I wanted some new challenge with which to redeem my cowardice. None came.

Soon I was standing beside Jack. The big Missourian's attention was fixed on the center of excitement. A large iron cage had apparently been dropped from the back of a wagon.

In the cage was an old black bear. I had never seen such an animal outside of a zoo, and admittedly this specimen did not impress me as much as the charging buffalo bulls I had seen. It was large enough, but mournful looking, as though its day had passed. One old timer declared he had not seen a live bear in this part of the country for nearly twenty years.

Two swells from Kansas had caught the animal. They were a gawky pair of flush faced newcomers, a little wary of the crowd of Texans around them, but proud to have done something worth their attention. Foremost in my fuzzy mind was the question of where they had gotten the cage to hold the beast, but I was in no mood to try and make myself heard among these chattering fools.

I tugged Jack's arm.

"Let's go, Jack."

The Missourian was in a trance. He had drunk a good deal in my absence, and my simple tug proved it by setting him to wavering on his great fur clad feet like a ponderous old oak in a high wind. His dark eyes didn't flicker from the shining, red lined eyes of the animal.

The look on his face was one of sad empathy. Man and beast had met through the cross hatch of rusty iron and achieved some primal state of recognition. Jack cocked his head slightly, and the bear mimicked the motion.

"You're drunk, Jack," I mumbled.

Jack nodded, but I did not think it was in answer to my observation.

"What you gonna do with that son of a bitch?" someone in the crowd asked of the Jayhawkers.

This provoked a short moment of perplexed quiet before some joker stepped up and offered to get the bear drunk. The motion was quickly passed, along with a hat for contributions to the bear's drinking money. This had to be done twice because someone made off with the hat the first time.

The second run to the saloon with the collection hat was made (under armed escort this time), and while the selected agents fulfilled their duty, someone pushed a mug of beer through the cage and tilted it downwards, showering the bear's sullen head in hops.

The bear groaned, and I pitied it. Beside me, Jack stirred uncomfortably.

The bear shifted in the cage and brought its great leathery paws up against the bars, sending a rattling reverberation down the length that caused us one and all (except Jack) to instinctively shrink back.

"Look at them claws," someone observed.

"Shred a man shore."

The bear lapped at the beer, and moaned and shook the cage when its owner pulled the mug away.

Someone suggested siccing 'Jinglebob's dawg' on it, and all present who were familiar with the canine agreed it would be a good fight. Odds were laid down, but when the beer bottles arrived, it was decided to forego the gladiatorial event in favor of giving the bear one last row before pitting it against the hound.

"They ought not to do that," Jack muttered, as the beer began to flow. I did not know if he meant the proposed fight or the current display. I was in agreement however. To see my first bear in such a state left me with a feeling akin to the sight of my first buffalo bull dead in a ditch.

The smell of bear and beer intermixed in my nostrils and churned my upset stomach. I tugged Jack again, more forcefully.

"Let's get outta here, Jack."

"Awright," Jack agreed, and reeled after me.

We made our way back to our esteemed lodgings and found Roam out front waiting for us, his saddlebags over his shoulder. Even in the dim light of the lamps and through my inebriated haze, I could tell something was wrong.

"Fuke with you?" he asked, as we came near.

"No," I answered, straining to see his dark face in the night. "Where is..."

He cut me off, and his news was as sharp and unwelcome as the taste in my mouth.

"Boys," he said. "We's flat busted."

Seventeenth

"No Cash, No Checks," Roam said. "I done tore through every inch of the wagon and every kit twice."

"What're you talking about?" I said. "The poor box was in his saddlebags, right under his head!"

"Them bags was empty, Stretch," Roam said gravely. "The box weren't in it."

"Mebbe one of them Texas boys smouched it?" Jack suggested.

"All's I know is it ain't there now."

"Mebbe somebody grabbed it by mistake?" Jack said hopefully.

"Yeah, and maybe they goin' give it right back if we explain it to 'em," Roam smirked. "Couple thousand dollars in gold and notes in that box."

"Well *you* had it last!" I exploded.

Roam's eyes cut into me, but he said nothing.

"Whalp. Ain't no use in pointin' fingers, I guess," Jack mumbled.

"Where Fuke at?" Roam asked Jack.

"He's at a whorehouse up the street with that Pat Garret fellow," I answered.

Roam sighed.

"How much you two got twixt you?"

I had very little. War Bag had not believed in advancing us for anything but necessities, and Jack had been fleeced by Fuke more than once over the past few weeks.

"I got some cash money left from the mule, and what the Mexicans give me. I'll split it 'mongst you, but don't tell Fuke."

"T'aint hardly fair to cut him out, is it?" Jack said.

"He got enough of your money to last him, Fats," said Roam. "Let him cut loose a little tonight, we'll drop it on him in the morning."

Roam got some cash out of his saddlebags. There was about eighteen dollars.

"There's more, but I gots to keep it to pay the sawbones for seein' to the ole man," Roam explained. "This ought to keep a roof over your heads for a few days, and put some food in your bellies till we can scrape together some more. I 'spect I can sell off the wagon and the bull team."

"Whar's War Bag?" Jack asked.

"I got him put up at a friend's place till he back on his feet," Roam said. He dished out the money to us. Nine dollars a piece in coin. "You tell the hotel-man, maybe he give you the money for the ole man's room back."

Maybe the company of the other white men had infected me somehow. I found myself asking,

"Where are *you* staying?"

"Same place as the ole man," Roam said. "Ought to go and see 'bout him as it is, afore it gets too late to find my way."

"Maybe we should come with you?" I suggested.

Roam looked at me, his mouth thin.

"Best not," he said. "I'll come lookin' for you roundabout noon tomorrow. Y'all get some sleep."

"Awright, Roam," Jack said. Earnings be damned, sleep was Jack's immediate need. He lumbered into the frame structure, ducking as he went.

Roam turned to leave.

"Do you think he'll live?" I asked. "War Bag?"

"Dunno," said Roam. "He ain't woke up yet."

I nodded, and bid goodnight. I went to the doorway, and paused to watch Roam go. I was entertaining thoughts of following him. What if he was absconding with our money?

Then out of the darkness of the street came Jinglebob and his two friends from the Bee Hive. The lamps inside had been turned low to accommodate the early sleepers, so I was invisible in the shadow of the doorway. The three men affixed their hateful eyes on Roam.

"What you doin' out so late, boy?" said the one with the unshaven neck. His eyes, bright as nickels, were utterly without the light of higher reason. The whites around them were cracked. He was like a wild animal somehow given the power of speech.

I saw Roam's shoulders go rigid. He didn't answer.

Jinglebob stared at Roam with naked contempt.

At his sides, Roam's fingers worked as of their own accord.

"You from the fort, boy?" asked the tall, handsome man in a deep southern drawl. He had a friendly smile on his ruddy face, as though he would work his boot heel into a festering wound and expect you to take it as a 'how do you do'.

"If he's from the fort, wheah's his bloddy stroipes?" Jinglebob observed.

I leaned against the door frame, peering from the shadows. The pit of my stomach was hot again, but not with spirits.

"Maybe he's a deserter," suggested the tall man, as if they were all talking about someone other than the man standing before them.

Roam's hands shrunk into fists. Even in the flickering lamp light, I could see the knuckles were tight under his dark skin.

"Nah, he's a ditch digger," the one with the unshaven neck said, looking from his companions to Roam. "All them nigger soldiers know their way around a shovel. "Even them blue bellies know there ain't no ditch digger like a nigger ditch digger. Want to come'n work for me, boy?"

My mouth went dry. They were pushing for a fight for no reason other than they wanted it. There was an unreasoning hatred in their faces, and something else almost too dark for me to imagine.

"Thet's an Ahmy pistol you gawt, ain't it?" Jinglebob asked, nodding to Roam's pistol belt, causing his ear to jiggle anew.

Roam looked down at the Remington. He made no move to touch it.

"Be a good 'lil wog and let's see it," the albino said.

The tall man smiled, but his eyes were like a slap in the face. The unshaven man kept on staring, chewing his chapped lips.

Somewhere a woman cackled hideously, and a dog barked at the night. Behind me I heard Jack calling for Whisper lowly. A man somewhere inside was singing in another language.

The tall man slapped Jinglebob on the back, sending his ear lobe into a paroxysm. It was a startling gesture that should have gotten them all shot. Maybe he had meant to set Roam off. I was the only one who jumped, and no one saw me.

"Aw, come on, Jinglebob," the tall man said. "Let's leave this nigger be. He's a good ol' boy."

"Hold on a minnit, Larn. I wanna see his pistol," Jinglebob said gravely.

The unshaven man spoke, his face cracking into a wide, unnerving smile.

"Well I wanna see my Minnie," he announced.

The tall man nodded in agreement.

"You're too damn drunk to shoot anyhow."

With that he curled his arm around the albino's neck and pulled him away like a school chum. The unshaven man lingered for a moment, as if daring Roam to shoot him in the back. Then he turned followed the others.

I waited to see if Roam would do it. He did not turn his head to watch them go, just kept staring ahead, still and silent.

The tall man was calling to the unshaven one before he had scarcely walked away.

"Edna's gonna find out about your Minnie one of these days, you old Tomcat. Then what're you gonna do?"

"What's *she* gonna do, you mean?" was the unshaven man's answer.

Their laughter faded between the buildings like the banter of goblins.

Roam stood there in the pool of flickering light cast by the dim lanterns of the Bison. Then he quietly resumed his course. His boots resounded on the planks until they stepped off into the muddy street and he was less than a departing shadow in the black.

I went inside.

I found Jack stooping down calling for his cat. I did not mention what had happened.

As we lay down on our potato sacks and blew out the light, I said to Jack,

"What do you suppose happened to our stakes?"

"Cain't be shore."

"Do you really think those Texans took it?"

"My paw tol' me a thief will allays look into his cup afore he dranks. Ah din't pay no mind to any of them fellers whilst they was drankin' though."

"Jack?"

"Mm?"

"Do you suppose . . . do you think Roam is . . . do you think he's telling the truth, about the money?"

"Ayuh," Jack said, rolling over, his voice dimmer as it faced away. "Why wouldn't he?"

"Well," I ruminated, ready to voice the grand suspicion I had been nurturing. "Why is he so keen on staying away from us?"

"Ain't got no choice, I 'reckin. Din't see no nigger hotels when we rode in."

"Well, but..." I thought about it. "Do you trust him?"

I heard Jack shift on the floor in the dark.

"Ah reckin thar ain't no lyin' in him."

I drifted off to sleep ashamed of my ill thought, and of my inaction in Roam's defense. I did not truly believe the three ruffians had meant to kill Roam, but I had said nothing, and a word might have been all the situation would have called for. Would I have suspected Fuke of hoarding the money? Would I have come to his defense?

I had another nightmare. Something about a red painted man with trembling arrows like the stalks of an ant's feelers instead of eyes, pulling a cart that was piled high with burning bodies. The cart was covered with etchings, and I was trying desperately to read them, but they were gibberish. I felt that if I could just make out the ciphers, I would avert some terrible disaster . . . .

We met Jim and Windy Bill at the lunch counter the next morning. Jim was sleepily dishing himself a plate of buffalo meat while his partner elbowed another skinner for cutting rights to the hunk of bread at the end.

"Morning, boys," Jim mumbled, and took his plate outside.

"What's the good word?" Windy Bill asked.

"Breakfast, I hope," I said, taking up a tin dish and shrugging my way between two men eating over the counter.

"You'll have to find a better one," Windy said, smiling and nodding to the bin of shredded meat, which was blood red in places. "The cook hasn't figured out how best to kill the buff before he tosses it into the pot. I s'pose he thinks if the fire doesn't do it in, the dogs he mixes in will."

"Dogs?" I asked dubiously.

"Sure. He's got a deal with Jinglebob Beddoe, the fella they pay to shoot strays. The owner invests in the bullets and gets his return in dog meat."

"That's a damn lie," muttered the skinner next to Windy. He crammed a piece of bread between his lips and carted his plate of meat away.

Windy whispered to me conspiratorially.

"Owner's cousin."

The men around the counter shook their heads and chuckled. I decided to entertain safety and stuck to the bread.

I left Jack ladling a mountain of meat and followed Windy out to the front walk. Jim was eating his breakfast and watching the early morning traffic pass by. We sat down on the edge of the boards with our boots in the street and watched as the butcher opened his shop across the way. He swept broken glass off his front walk into a tin dustbin, and clouds of ruddy dust boiled around his feet.

"Yep," Windy remarked. "Jinglebob hands the strays over and old Tubry processes 'em. Goddamn model of modern efficiency."

"This Jinglebob," I said. "Is he an albino with a ragged ear?"

"There's no other I know of," Windy admitted. "You met?"

"Yes," I said, rubbing my temples and thinking sadly of the hurdy gurdy gal from the Bee Hive and Roam's encounter of the previous night.

"His story's more colorful than his flesh, I'll tell you. He's a New Zealander. They say he cut a Chinaman to little pieces outside a Barbary house in San Francisco."

"For what?"

Windy shrugged.

"For being a Chinaman. Now the town fathers pay him twenty five cents a tail to keep the strays out of the hide yards."

Jim stood up, wiping the grease from his lips with the back of his hand. His plate was empty, and he moved past us to tote it back inside.

"You talk too damn much, Windy. You know that?" Jim said in parting.

As Jim went inside, Jack lumbered out. We watched him peer about the corners of the frame lodge for Whisper, who had not made an appearance yet.

"He'll turn up, Jack" I offered.

"Ah cain't figger whar he's gone off to," Jack said, as he settled down between us.

"Cats will wiggle their way into the smallest places," I said. "He's probably found a nest of mice over in the hide yard."

"Better hope he hasn't," Windy said. "Once in awhile Jinglebob will catch a dog for the fights, but cats he kills on sight. The only four legged critter smaller than a horse that has free rein around here is Jinglebob's own mutt. That's the nastiest dog I've ever seen. Ugly sonofabitch with a chawed up ear like his master, and one crazy rollin' eye. You know how they say people start to look like their pets? Well, there you go. Say, I heard someone caught a bear and that they're going to sic that hound on him."

"It's true," I affirmed. "Jack and I saw it last night. They were giving it beer."

"That'll be a sight. Jinglebob's dog has come out on top in the last eight fights. But a bear! I wonder what odds they're giving over at the Bee Hive . . ."

"That's where the bear is," I said. "Out front. Two men have it in an iron cage."

"Well! Where did they get a cage big enough to hold a bear?"

I shrugged.

A group of Tonkawas, shabby things in drab clothes with unwashed hair, came up pulling a cart of firewood for the hotel's proprietor. I was briefly reminded of my dream.

Windy put his plate down and stood up.

"Think I'll go over and see about it. I'll see you boys later."

We nodded our goodbyes and he was off down the avenue to see the sight.

Jim came over and sat down on the porch and began rolling a cigarette.

"Your partner's off to see the bear," I said.

"Never in my life have I heard a man who talked a longer blue streak than him," said Jim.

He offered me the cigarette he'd rolled, and I thanked him and took it. The smoke was hot in my lungs, and agitated my cough so that I would have preferred to mash it out. In the end I thought it might seem rude.

"Jack told me you got jumped by Injuns out there."

I nodded, thinking of the rain water beading on the back of the brave I'd killed.

"Me and Windy are plannin' on joinin' up with Wright Mooar's outfit. They're pullin' out on New Years, maybe sooner if the snow don't hit too hard. Might be they could use a few men that got some experience fightin' Injuns. Lord knows I don't know a thing about it, and Windy'd ruther flap his gums at a painted savage than shoot him. You interested?"

I shrugged.

"I don't know, Jim."

"Could use another skinner too, if anybody else from your outfit needs the work. Word is they're gonna concentrate on hides. It's where the money is getting to be. I hear tell your boss is laid up. Lost an arm?"

"Eye."

Jim nodded and smoked.

Jack looked worried, and was wolfing down his food.

"What's the matter, Jack?"

Jack was standing up as I finished. I had never once seen him leave a plate so slovenly polished.

"Reckin' I'll go an' check the hide yard fer Whisper."

"Roam's gonna be looking for us."

"Not till noon. Ah reckin Ah'll be back by then."

There was no stopping him. The Tonks cleared him a path.

The stale bread worked its absorbing powers on my alcohol-ridden body. The pains in my head soon departed before the cool breeze that made its way down the Avenue, toward the great hide piles that it seemed, with all the planning in the air, would only grow greater. Not that I could smell their stink anyway. I sniffed, and it was a noisy performance. I needed a bath.

Up the street came Fuke, looking content. He wore a self contented expression beneath his mustache, and his capote was over his shoulder. He seemed to suffer no ill effects from his sinful night. His hair was neatly combed and his skin clean. His face hair was trimmed and a new white shirt clung to him beneath his suspenders.

"*Mornin'* glories," he said, and placed one foot on the boardwalk between Jim and me.

"You look pleased with yourself," I said.

"Well, I accomplished a *helluva* lot last night. Now just where is that *nigra* and our cash?"

I looked up at him. He had not even inquired about War Bag.

"He's around. He had some business to attend to."

Fuke sat down and wrinkled his nose at Jack's plate.

"Is this the *slop* they're feeding you boys here? You'd have done better to stay at *Sally's*. It was bacon and *eggs* from the chicken yard this morning. Barney brewed chicory, and Julietta hustled on down to the river and scrubbed our clothes. Not bad for the price."

"Sounds like a charming place," I muttered, hearing Barney's laughter from the night before like a hollow echo.

"Not without charm, *surely*," Fuke admitted. "Why not come with me tonight, *Juniper?*"

"I don't think your new pal Pat likes me."

"I suspect it's the other way *around*. Come on. You've been staring at ah . . . buffalo *hocks* too long. Or are you still pining over that wench from Colorado?" He mimed a pondering expression, scratching his furred chin reflectively. "Now what *was* her name again?"

In spite of myself, my ears colored.

"You know well what it was. You probably paid your Mexican whore to answer to it last night, you damned profligate."

Fuke smirked.

"It was so hard for her to *pronounce*, too. But her mouth was a bit too *busy* to bother with ah . . . *proper* articulation."

I shook my head at his talk.

Fuke laughed and slapped me on the back. He stood up and stretched, his joints popping in succession.

"Come on, *Juniper*. We'll go and and get you a *proper* breakfast at least. Maybe you'll feel more up to the *task* once you've got some decent food in you."

"I don't have the money for it."

"That's alright, you poor *bastard*, I'll stake you."

You can't afford it either, I wanted to say. It was obvious that Fuke was spending his money like water and it couldn't possibly last him much longer. How would he react when he found out there was none waiting for him either?

"How about you?" he asked Jim.

"Aw, I reckon I ate my fill."

"Well, you come and find us in a half *hour,* if you want. I imagine the pig'll have worked its way through the ah . . . *python* by then."

Fuke pulled me to my feet and led me off down the street.

"I'm really not hungry, Fuke," I said.

"Well, I didn't intend on staking you *anyway*. I want to talk to you about what you plan on doing once we get our *allowances.*"

"Maybe I'll ride up to Kansas and join up with a cattle outfit," I mused.

Fuke looked at me.

"*Cattle?* Working cattle's a fool's trade. It's no way to get ahead, *Juniper.*"

"That's easy for you to say. You're a shooter. I don't see much of a future for myself in skinning."

"Now there's the *first* bright thing I've heard you say."

"So what do you intend to do, then?"

"Well, I believe I've found my new occupation, *Juniper.*"

He guided me to The Bee Hive Saloon, and led me up to the porch. Across the street, the crowd around the bear had increased. Someone had paid a Tonk boy his weight in pennies to stand there with a stick and prod the animal through the bars of its prison. Its miserable sleepless groans filled the street. I pitied the animal, but I had learned my lesson with Cutter Sharpes' mule. Anyway, popular opinion outweighed any humanitarian notions I harbored.

Fuke snapped his fingers to get my attention.

I looked dubiously at our destination.

"You've decided to be a professional drunkard?"

"*Ah*, if only they *would* pay me for it" he smiled, and tapped a signboard nailed to the wall.

It read, FARO DEALER WANTED. INQUIRE WITHIN.

Eighteenth

"*Faro?*"

When I asked Fuke why in the world he wanted to be a faro dealer, he answered,

"Why do anything, Juniper? Besides, an *Englishman* offered me a job with a big buff running *outfit* that's pulling out at the end of the year. *Shooting,* of course. Might as well spend the next few *months* doing something *interesting.*"

Fuke walked boldly into the Bee Hive with the Faro Dealer advertisement tucked under his arm, and me alongside. It was too early for the place to be full of anyone other than the most unrepentant drunks. One red eyed bartender, a Negro sweeper who looked to be somewhat addle-minded, and two or three tired looking girls were all the staff on hand. The band stand was empty and the place was a forest of unpainted chair legs sprouting from the tables that the Negro with the broom negotiated in a slow and practiced shuffle. Broken glass clinked now and then, and the Negro's straw broom sifting through the sawdust was like the sound of an old man walking his cellar.

Fuke took notice of the girls just as I did. There were three of them, in the black and yellow show costumes. There was a chubby, pale red head with one black leg resting on an empty chair. Beside her was a terribly tall and bored looking Mexican girl twisting her thin hair. My heart did a turnaround in my chest when I saw that the third was my Cathy.

We both took off our hats, and smiled their way. While two of the girls sipped from caramel colored bottles and looked plainly disinterested, I saw Cathy smile pleasantly. My spine seemed to hum like a telegraph wire.

"Boys thirsty?" the barman called to us.

"Not just *yet,*" Fuke said back, and went over to the bar. I followed, not knowing what else to do. I was conscious of the girls watching us.

Fuke clapped the placard down on the counter top meaningfully.

"You boys ever dealt faro before?" the barman asked in a tired voice that reeked of sour mash.

"Never even seen it played," I said.

Fuke glanced at me, then smiled at the barman. He was spinning his wide brimmed hat on his finger.

"Speak for yourself, *Juniper*," he said. "*I* know the game *soda* to hoc."

The barman ignored me from then on, and looked to Fuke.

"How many turns in a game?"

"Twenty five," Fuke answered right away.

The barman nodded to himself, and reached under the counter. He produced a small box with a fanciful but marred orange tiger painted on top. He pushed it across the scored bar top to Fuke.

"Faro table's straight back thataway," the barman said, waving his hand in the general direction. "Go over to them girls and ask for Cathy. She's lookout at the table most nights. You can run a practice game, and if you can't cut it, she'll tell me."

"What's the ah...*wages* for this job, professor?"

"You ask me after you've dealt a couple through."

Fuke nodded and scooped the tiger box off the bar. He went across the room to the table of women, with me dragging my heels beside him.

"Just keep quiet now, *Juniper*," he whispered to me. "I won't have you putting my *chances* in jeopardy."

"Chances for what?" I whispered back.

But he flipped his hat onto a set of antlers mounted on a post and turned his eyes on the girls.

It didn't take long for him to wrangle them over to the faro table and get a game going. He bought them real drinks, and soon they were happy to correct his mistakes and get one of his polite 'well now thank you my *dears*' or a winsome wink of his blue sky eyes in exchange.

Even Cathy was under his spell. As for me, I went unnoticed and worse, unrecognized. Of course, I didn't say a word to her. I suppose I was rendered as speechless by Fuke's machinations as they were wooed. I sat back on a stool and concentrated on the chipped table with its layout of card pictures rubbed faint by the elbows and coins of many a hopeful gambler (or punt, as they called them). How many hard won salaries had been lost on this table?

The cards flitted from the tiger box and in the time it would have taken me to master the basics of play, Fuke was flipping them about and walking the Chinese copper across his knuckles with all the deftness of a veteran sharp. It annoyed me how easily he seemed to take to some things, and how easily people took to him. I consoled myself by remembering how he'd stuttered through *Titus Andronicus*.

I rested my chin on the back of my chair and sipped warm beer, the giggling of the inebriated girls in my ears as they began to bet articles of clothing on the cards. Cathy's laugh was the shrillest of all.

I watched her move her lookout's stool closer and closer to Fuke, and soon he was resting his hand on her black stockinged knee.

The barman kept the drinks coming, encouraging Fuke. The dimwit sweeper paused in his work to watch from a corner as the red head and the tall Mexican girl began to lose.

Soon the red head's flour white breasts were resting on the green faro table. She had let her rust colored hair down to obscure her nipples, which were like peach colored thumbs. The gangly Mexican girl was laughing a wind-less, shaking laugh as she stepped out of her bloomers, struggling to pull her corset down to cover her lower regions. Her large breasts kept threatening to burst forth. Cathy was fogging a whiskey glass and making eyes at Fuke. He was ordering a fourth round of drinks and I had just about resolved to gather up what was left of my respect for womanhood before Fuke dashed it utterly. Then there was the sound of a commotion outside, and of feet hurrying along the boardwalk.

A breathless cowboy pushed open the door and spluttered.

"Some crazy sonofabitch is strangling Jinglebob's dog right in the street!"

Fuke and the girls paused in their revelry, but I wasted no time in pushing out of my chair and ducking past the bare bottomed Mexican girl. Better to see a ruckus than sit by and watch Fuke seduce my hurdy gurdy gal.

I jammed my hat on my fat head as I went to the front porch to join the spectators. Pat Garrett was standing there already, and when he saw me, he said,

"Ain't that big fellow with you?"

Jack was standing straddle-legged in the street with a crowd pressed around him. He had the neck of a mottle colored dog in-between his two huge hands, and was throttling it harshly. I was stunned, but saw right off what he was after.

From the snapping jaws of the newsprint colored hound tumbled Whisper, a little frazzled and soaked with drool, but alive. The cripple cat rolled once in the dust as he fell from the dog's maw, then skinned out like a gust of gray wind, streaking in-between the very legs of Jinglebob as he stalked angrily up the street, one hand on his pistol.

The New Zealander was cussing up a storm in his peculiar dialect.

Jack had the dog partway off the ground, and it was whining terribly (in between slavering barks), its hind legs scrabbling at the dust.

Jinglebob stopped within two feet of Jack and drew his pistol, screaming at Jack to leave off his dog.

Jack turned with a speed I couldn't guess he had and batted Jinglebob's pistol from his hand as though he were pushing aside an unwelcome bouquet of pansies. Jinglebob grabbed his knife.

Jack let the maddened dog fall and grabbed at Jinglebob with both hands, pinning his arms to his sides.

In answer, Jinglebob threw back his head and drove it straight into Jack's face, smashing his lumpish nose with a crunch. This sent the Missourian into a berserk. Jack dropped one fist from his shoulder and brought it up into the New Zealander's bread wallet with such force that it lifted him clean from the street. Continuing the motion with that one hand, Jack tossed Jinglebob head over heels with a splash into the nearby horse trough which, not having been designed to serve much beyond a utilitarian purpose, smashed to pieces and soaked the feet of the lookers-on with dingy water.

Jinglebob's mutt shook its head as it got to its feet. Seeing its master so mishandled, it leapt at Jack's bandaged arm, tearing into his sleeve and wrappings with an unmatched ferocity. Jack's sleeve ripped open and I saw his red blood break out in the sun as the dog dragged its fangs through his arrow wound. Jack reached down with his free arm and gripped the dog by the skin of its rump, while twisting his trapped hand to latch onto the nape of its neck. Thus he lifted the dog completely off the ground and flung it from him, right at the crowd of gawkers.

The citizenry fled in every direction from that flying, growling mass of ire and tics. The dog crashed to the ground and rolled on its back.

Jinglebob sat up dazedly from the broken water trough and felt for his knife again.

At the front of the Bee Hive, Fuke, the girls, the barman, Pat Garrett and I all stood stricken in amazement. Across the street, I noticed a familiar face come out of the dry goods store. It was one of the other men that had accosted Roam the night before in front of our hotel. The tall one, whom Jinglebob had called Larn. I tensed, thinking he might try and intervene. But he just stood there, watching with all the rest, smiling a little.

Jack paused and wiped at his streaming nose. He succeeded only in smearing more blood from his arm across his face.

Jinglebob's big-shouldered hound scrabbled to its feet and went charging at Jack again. Ugly and mean as it was, it was not at all bright.

Jack crouched to receive the mad dog, his bloodied hands curled into fists. As the animal reached him, Jinglebob, on his knees, lunged at Jack's flank, his knife in hand.

A cry went up amongst all who saw the blade of the big bowie disappear into the Missourian's hunched thigh. Jack seemed to take no notice though. He drew his big fist back and met the snarling dog straight on. The dog seemed to collapse in on itself like a concertina as its snout collided with Jack's ham hock fist. It fell stunned to the dust, whining and kicking.

Jinglebob retracted his knife and the splatter of blood that issued from Jack's leg seemed to awaken the Missourian to his predicament. He turned his attention on the surly little albino, and caught the knife hand as it swung towards him again. His other fist slammed against Jinglebob's head, and the New Zealander's skull lolled on its shoulders like a sprung jack-in-the-box. Jinglebob's legs melted under him and he fell on his back in the dust.

But his tenacious hound was not satisfied, and sprung full at Jack's face, burying its muzzle in the spot between his neck and hefty shoulder like a wolf trying to bring down a moose. Jack tilted his head back and his face registered agony. His big fingers found purchase in the animal's furry hide and clamped once again around its neck.

There was a moment of struggle where the Missourian's hands curled in the dog's filthy fur like the desperate talons of a blood-mad gorilla. The dog released its grip and the Missourian flung it down with malice. Not in the mood to offer the dog another chance, he took hold of the hind legs.

Jack rose and swung the dog up and around in a wide arc that ended abruptly when it met the hard wood of the support holding up the awning of the barber shop. There was an audible break when the dog's head rebounded off the support. The animal gave a heart wrenching cry and then landed for the last time in the dust, limp and motionless.

Jack fell to his knees and saved himself from eating dirt with the palms of his hands. Blood spattered the mud in dark little dots.

A man with a walrus-style mustache came tromping up the street, a shotgun in his hand. Two men in suits and wide brimmed hats, one with spectacles, trailed behind. The man with the shotgun muscled his way past the citizenry and stood over Jack, who was slowly getting to his feet.

The armed man seemed to allow three seconds to the scene before him, before making his decision. When I saw what it was, I called out from the porch of the Bee Hive,

"Jack!"

The big double barrel of the shotgun swung down hard against the side of Jack's head, and the skinner's eyes rolled up on him. He fell face first into the mud.

The two men in suits rushed past us, wringing their hands. The spectacled man dabbed at his forehead with a blue hanky.

"Dammit! Goddammit!" I heard him say.

The man called Larn stepped out of the doorway of Jackson's store.

I rushed out into the street.

I saw the man with the shotgun turn to Jinglebob, who was trying diligently to hoist himself up on one elbow.

"Beddoe," he muttered. "I shoulda known."

He turned and looked at me as I knelt down to see to Jack. He had a deputy's badge pinned to his coat.

Jack was breathing, but he had a fresh wound on his scalp where the end of the barrel had ripped his flesh.

Jinglebob crawled over to his dog, which was lying limp and forever disinterested in further proceedings.

"You could've stoved in his skull," I admonished the deputy.

"Not likely," he quipped.

"He needs a doctor," I said.

The deputy regarded me for a moment before the two men in suits crossed the street towards him.

"Reddick! What the hell is this?" hollered the spectacled man.

"Shit," Deputy Reddick said to himself as they approached.

"I'd say," said the other man, having caught the utterance. "How in hell do these people expect to make this hole the county seat with such . . . such . . . occurrences." He turned suddenly to the people in the street and implored them with a dramatic sweep of his trembling hand. "How in the hell do you people expect to win the county seat with such...barbarity running rampant right here in the street?" He seemed more pleased with the revision.

The bear gave a plaintive moan then, startling the man in the suit and nearly everyone else. He poked his nose through the cage and snuffled, wondering where his audience had gone.

"What the hell is that?" the man in the suit demanded, pointing at the bear. "Reddick! What the hell is that?"

"It's a bear," Reddick answered, tilting his hat back on his head and peering in amazement at the caged bear.

"Well what in the hell is it doing in the street?"

"Hey Cornelius, want us to put him up at your place?" someone in the crowd yelled. Others burst out laughing.

The man in the suit went red.

Jinglebob cradled his dog and glowered hatefully at Jack.

I tugged the deputy's coat.

"He needs a doctor," I repeated.

"He'll get tended to once he's in custody," the deputy assured me, looking past the men in suits, at Jinglebob's tall friend, who was at the edge of the crowd looking on with interest. "Larn, you get that albino out of the street!"

Larn nodded and touched his hat.

"Sure thing, Deputy," he called.

"Pat!" Reddick called, as the Alabaman started to duck back inside the Bee Hive. "And you!" he shouted to Fuke. "Come over here." He looked down at me. "And you too, boy. Come on and help."

"Help what?"

"Help me carry this big bastard."

"Carry him where?"

"To Shaugnassey's, up the street. The calaboose ain't finished yet."

The bear gave another groan, and the man in the suit, who had stepped closer to investigate, thought better of it. It got quite a laugh from those who had been standing around.

As we hefted Jack off to Shaugnassey's, I heard the renewed debate as to what they'd do with the bear now that Jinglebob's dog was dead.

Fuke grunted as he hoisted his share of Jack.

"*Damn*, Fats. Just like a goddamn red *Indian*."

Nineteenth

Jack's Durance was not welcomed by his warden, Dick Shaugnassey.

"Come on now, Cade. You're not bringing him in here!" Shaugnassey called from behind the bar as we came across the hard packed threshold. He was a flinty eyed man with the look and build of a veteran pugilist.

"Sorry, Dick," Reddick grunted under the tremendous burden of the unconscious Missourian. "I'll just chain him to the cot in back like usual."

The drinkers at the bar made way for our little procession and watched us with interest.

"That's one big son of a bitch," one commented. "What happened to him?"

"He fought with Jinglebob's dog," Deputy Reddick grunted.

"Don't hardly seem like a dog could do all that to a big son of a bitch like him."

"Well, Jinglebob had a hand in it."

"Jinglebob dead?"

"Nobody's dead," Reddick grunted as we dragged Jack through the curtained doorway in the back. "He's down the street with Larn."

Shaugnassey held open the curtain as we got Jack into the back room, where a few barrels of whiskey and beer were stacked and a raw Army cot sat in the middle, next to an oil lamp on the floor.

The cot sagged with Jack's weight, and his big feet dangled over the end.

Reddick hunkered down and produced a set of iron manacles which he fitted to Jack's wrist and the frame of the cot.

"He'd better not try and break into my whiskey like the last one you locked up in there," Shaugnassey said from the doorway, his tenpin arms folded disapprovingly across his apron. There was a thick two foot piece of firewood with a carved handle in one of his fists.

"Just bust him in the head with that horse cock if he gets out of line," Reddick said, stepping back to inspect the fetters. "Pat, I'd like you to run up to the post and see if the doctor will come down and see this man," Reddick said over his shoulder. "I've no intention of bearing him up the hill till I can secure a wagon."

"I can do that," Pat said. "I thought Cornelius had set those colored boys to build a jail."

"Cheap bastard had 'em makin' it outta sod. I had to tell 'em to start over. What use is a jail if you can dig your way out?" Reddick said wearily. He leaned against the wall and rolled a cigarette. "They been upriver gatherin' stone for the past couple of days now."

"What about Jinglebob?" I asked. There was surely no room in this makeshift prison to hold the both of them.

"Well," Reddick said, dragging on his cigarette, "as I hear it, it was this feller that whupped the albino's dog. Beddoe was just protecting his property. Still, we can't rightly have him loose after pulling that knife like he did. I'll remand him to Larn's custody. We'll see if Beddoe wants to press charges."

"Is that fair?" I asked.

"No, I guess it ain't. But if I locked up that albino, Larn and his 'vigilance brothers'd' just come with torches and take him out. The Marshal is off chasin' them goddamned horse rustlin' Comanches with a posse of Rangers. John Larn is the closest thing there is to another deputy now since Bill Hendrix run off to Californee with that sportin' woman last week."

"*He's* the only law? *That* bully?" I asked, incredulous.

"Unless *you* want the job," Reddick said hopefully, striking a match on the door frame (to Dick Shaugnassey's disapproval) and puffing smoke.

I shook my head in disgust.

"How about you two?" he said to Pat and Fuke. I could see he was serious.

Fuke just laughed.

"No," Pat said. "I don't think so."

"Well shit," Reddick said. "I'm up the creek without a paddle here, boys. Leastways until the posse gets back."

"What about the Army?"

"Jurisdiction," Reddick shrugged. "That's what Major Schwann always says, anyhow."

"How long will Jack have to stay here?" I asked.

Reddick shrugged.

"Depends on what the albino says when he comes around. If he decides to file a complaint it could be a month or more before he sees a judge."

"You won't be keepin' him here for a whole month, I'll tell you that for nothing," Shaugnassey growled, scrubbing at the mark Reddick's match had made on the door frame.

Reddick sighed heavily and followed him out.

I looked down at Jack. He stirred and moaned. The blood was caking on his wounds.

I walked out to the bar with the others.

"Then again," Reddick was saying, "the vigilantes don't take to a slight against their own. Not even a no good dog killer like Jinglebob. You boys ought to get in good with Larn or his buddy Selman, maybe they'll keep Beddoe cool. I'll sit here with your man, but I'm just one fella. If they come for him, they'll have him."

I bit my lip. Reddick took notice, not without satisfaction.

"Mr. Dowell and Mr. Stribling got a drawer full of tin stars if you want the job."

Shaugnassey leaned across the bar.

"Not exactly the sort of talk we like to hear from our only lawman."

Reddick shrugged and got out a wad of money.

"Gimme a beer."

Pat cleared his throat in the cuspidor.

"I believe I'll go up the hill and see about that doctor."

"Maybe you'll catch him sober," Reddick told him, as he set his hat on the bar. I could see the ring of dark sweat around the band.

"Why not stay, *Pat?*" Fuke piped up. "You go on up to the post, *Juniper*. We'll stay with Fats here, in case of trouble."

I narrowed my eyes at Fuke. I could see he just wanted a drink.

"Well, we can't just let the big dumb son of a *bitch* get shot down. And they *will* shoot him, Juniper. I haven't seen a tree in all this pitiful goddamn *country* that's tall enough, let alone *strong* enough to hoist Fat *Jack* off his feet."

"We got trees for him too," Reddick mumbled into his beer glass.

"No offence meant, but there's *nothing* in Texas to match a Louisiana Cypress. Why, you could string up six black cottonwood *blossoms* at a time and not snap a branch. I *know*. I've *seen* it."

I had no idea what a black cottonwood blossom was.

Fuke tossed his arm around my neck.

"I'll buy you a *drink* when you get back," he said.

Pat sat back down.

"You ought not to be drunk if you're expecting trouble," I said.

"Don't fret, *Juniper*. We got the *law* on our side. Who can stand *against*?"

*   *   *

Up at the post they told me the doctor was out making calls. I did not ask when he would be back. There was a postmaster, and I purchased an envelope, a leaf of paper and a pencil. I scratched off a quick letter to my parents in Chicago on the side of Conrad's store:

Mother and Father,

    Am alive and well. Please do not worry. You will hear more from me soon.

Juniper

I grimaced as I realized what I had signed, and did my best to scratch it out and write my name over it. I couldn't afford to waste more money on paper, and my notebook was back at the Bison. Besides, it had been a chore to wrestle through the crowd at Conrad's for the paper in the first place. They would surely know their own son's handwriting, I told myself. I sealed the envelope and delivered it to the postmaster.

It was a lonely walk down Government Hill as I thought of all that had befallen our little band, and of my parents back home. It seemed as though I were estranged from two families. I realized I couldn't clearly picture my father's face. I kept seeing War Bag. Thinking of the old man made me remember my appointment with Roam. The sun was well over the highpoint.

I hurried the rest of the way down and cut across the Avenue and along the butcher shop to the Bison. Roam was nowhere to be seen, and the taste of the envelope was bitter on my tongue.

I went back towards Shaugnassey's, and stopped when I remembered Whisper. The last place I'd seen the crippled cat run was under Jinglebob's legs in the middle of the Avenue, in front of the Bee Hive.

After all Jack had gone through for the cat, he would surely want to know that Whisper was alright. I went over to the Bee Hive and got down on my knee to peer under the boardwalk. The sun made clear bright intersections in the shadows beneath. Huddled there against the foundation was Whisper, blinking at me with an air of aloof superiority as though he were bragging to me how he had avoided the clash that had gotten his master arrested.

I called the cat lowly, but Whisper only blinked at me in a bland, noncommittal manner. I got down on my belly in the wet earth, and stretched my arms under the walk. Whisper tensed as my groping hands passed into his personal space. He actually had the nerve to hiss. I made kissing noises with my lips, and spoke in a cooing voice. I had to wriggle a bit in the mud, and duck my head and shoulders into the gap under the planks to gain the necessary reach. When the fingers of my left hand were within inches of his matted flank, the son of a bitch lashed out and painted my knuckles with a neat red slash.

I hit my head on the plank and cursed loudly.

"Stretch, that you?" came a familiar voice.

I pulled myself out from underneath the boardwalk and saw Roam sitting atop Crawfish. There was a spectacled, yellow bearded white man in a blue wool coat riding alongside him.

"I'll be off then, Welty," the white man said.

Roam saluted the man, I suppose out of long habit.

"Yassuh. Obliged to you, suh."

"Who's that?" I asked, standing up as Roam came over.

"Post sawbones. He come out and took a look at War Bag."

I sucked on my bloodied hand.

"I was just up at the fort looking for him. Jack's been injured."

"How'd *that* happen?" Roam said to me as we walked down the street.

"It's Jack. He killed Jinglebob Beddoe's dog. They've got him chained to a cot in back of Shaugnassey's Saloon."

I glanced at Roam, and noticed his eyes flicker at my mention of Jinglebob. He knew who I meant. He didn't know that I'd seen him last night accosted by Larn, Jinglebob, and the other man. I didn't mention it.

"How's War Bag?" I asked quickly.

"The doc says he'll live," Roam said. "Somethin' else."

"What?"

"War Bag's awake. Wants me to bring y'all to see him. 'Says it's got to do with the money."

Twentieth

The Colored Settlement lay just over Collins Creek. It was a community of about one or two dozen Negro families clustered together, contending with the threat of the red man on the one front, and living in dread of the whites on the other. This was evidenced by the rifle crosses cut into the shutters of every cabin, and the apprehension with which our arrival was met by the shy faces of all those we saw.

Colored women stopped their sweeping and scrubbing and went indoors at the sight of us. A pair of long limbed black boys ceased their game of mumbelty peg and watched us warily from beneath the unfathomable shadows of their wiry straw hats. As we passed, the taller boy folded up his jackknife and quickly put it in the pocket of his jeans, as though it were a treasure that we, like feudal lords, might demand that he relinquish. A little girl in corn rows peeked like a field mouse around the corner of a shack and did not return my smile.

I had never seen a community of blacks together. My skin felt stark and pale under their silent eyes. I was a trespasser who could go where I pleased, but was not welcome anywhere. They all shared the same dull, stoic look I'd seen Roam assume when Fuke had spoken to Pat Garret in his stead.

Fuke seemed to be making a concerted effort not to regard anyone around him. His eyes were straight ahead, looking even beyond me. Perhaps he was drunk.

Roam took us to a plain cabin off by itself with a patch of cultivated earth. There was a pen with a lone pig in it, and an axe buried in a stump by the porch. A pecan tree stood in the yard. The spindly branches were bare of leaf, and clawed the grey sky like petrified lightning. A squirrel had its nest up in the high branches, and he chattered at us as we rode up.

The cabin was in good shape for its age. It was one of the older ones by the weathered shingles and the sun-faded wood. The pig pen needed mending. There was a work shed around the corner of the cabin, and I could see a sawhorse through the half open door. Solomon was tethered outside.

Roam tied Crawfish to the stump. From the open door of the cabin, an old Negro shambled out, a thin shadow in overalls under the porch.

Roam stepped up to the old man and took off his hat. The old man acknowledged the deference with a slight nod, and looked past him at Fuke and I, standing at the edge of the property.

"You mean to say he brought him *here?*" Fuke whispered to me. "Why didn't he pay to put him up someplace *decent?*"

"It's awright," Roam said. I did not know if he meant for us to approach, or to reassure the old man. We walked up anyway.

"This here's Milt Sutton," Roam said to us. "Mister Sutton, that there's Stretch."

"Suh," Sutton said, bobbing his head at me. He was in his sixties, and his skin hung below his chin, which was flecked with a silver foam of hair. He was bald on top, and the sun shined on his pate as though on a dome of solid jet. His skin was so dark it was like the scorched hide of a steam engine. It made his eyes seem red and feverish in his face. His hands were knots of age, and there were wrinkles even above his lips, as though he were mummified.

"And that's Fuke," Roam said, gesturing to Fuke, who stood with his arms folded across his chest. He was swaying slightly where he stood, like a man anxious to be somewhere else.

"What say, boy?" Sutton said to him with a smile. His 'say' came out like 'shay.' He had no teeth.

Fuke was taken aback. His lips parted momentarily, then he regained himself, and doffed his hat, revealing a stern expression.

The old Negro's smile fell, but his eyes flashed brightly. He bobbed his head in the exact same way he had done with me.

"No offense meant, suh. These ole eyes o' mine been gettin' lazier every year. I 'spect it won't be long fo' the mawnin' come when I jest can't rouse 'em a'tall. Y'all come inside. Mr. Tyler's suppin' jest now."

Roam followed Sutton in, and Fuke and I went after.

An abrupt motion caught my eye as we entered. I was suprised to see Trigueño, Orfeo's young nephew. He was standing in the corner, a rocking chair bumping against the wall. He was as wild and desperate looking as ever. I noticed his long cibolero lance leaning in another corner of the room.

Roam already had his hands up to calm the boy as Fuke hissed,

"What in the hell's *he* doing here?"

"*Está bien. Está bien,*" Roam told the boy, and Trigueño slowly sat back down, his serape draped about him like the regalia of some feral princeling. "He showed up yesterday, and ain't left yet. He jest watches the ole man. Ain't no harm."

"He's *touched*. How do you know he won't decide to cut his goddamn *throat*?"

"He wouldn't do that," said a rumbling voice, that even low and laborious as it was, made my hackles rise. "His people left. He wanted to stay."

The cabin had two rooms. There was a kitchen and sitting room as the foyer and a bedroom adjoining. Through the open passage we saw War Bag, propped in bed. A blue kerchief was tied around his head, covering his missing eye, and his hair was unbound, flowing about his shoulders, giving him a savage look. He lay like a stricken Odin, his forked beard a little whiter than I remembered. He was clean, but still weak. He sipped a bowl of broth and watched us like a patient panther.

"He thinks he owes me," War Bag continued, "for stealin' him back from the Apaches when he was a baby. That's as near as I can figure."

Fuke and I went to his bedside like excited children, talking at once.

"Looks like you'll pull *through*, old man," Fuke said.

"I may at that," War Bag said. His voice was hoarse and quiet, and speaking seemed to tax him. "Where's Fatty? I told you I wanted all of 'em here."

Roam looked at me and started to answer.

"We couldn't find him," I cut in. "He went off looking for his cat." No need to fret War Bag unnecessarily.

"Goddamn hillbilly and his goddamn cat," War Bag snickered. He coughed, and the mirth went out of him. He was sick yet from the exposure. There was more than one blanket on him, and he wore a flannel nightshirt as well. His hat and coat were on a chair in the corner. Bullthrower leaned against the wall, secure in its cracked case.

"Well, what'd you call us out here for, *old man*? Ready to *divvy* up?" Fuke asked, clapping his hands together.

I looked knowingly at Roam. He looked like a tornado was about to hit us all.

"You're a sentimental son of a bitch, ain't you?" War Bag hissed. "I figured you'd run through your cash like a widow through her bitters. And I know the rest of you probably got the same thing on your mind, though you got more manners than Fuke, so you ain't asked. Been waitin' for me to come to, I s'pect."

"It's not like that at all," I offered.

"Sure it is," War Bag said, sipping his broth. The veins in his hands stood out like telegraph wires.

"Well, you went and told the *nigra* here to hold off on payin' us without your *say* so," Fuke said, "and his word goes a long way with the other two. You left us high and *dry*, old man."

Roam gritted his teeth.

"You goin' push that nigra shit too far, Fuke."

Fuke smiled derisively, and spoke without looking at Roam.

"Ease up, Roam. You been a good *boy*. Nobody's sayin'..."

Suddenly Roam had a fistful of Fuke's long locks in his fist. He jerked Fuke right off his feet and sent him down with a thud to the hard packed floor. He was straddling him in the next instant, and tore his knife from its scabbard. He would have brought it down had I not reached out and grabbed his wrist with both hands.

"You get offa me, you *black son of a bitch!*" Fuke howled. "You get *off!*"

I could smell Shaughnassey's beer on his breath and in his sweat. It exuded from his very pores.

"You ain't worth a pig's shit without yo trigger finger and 'yo cracker freinds around, you goddamn *swell!*" Roam hollered. It was as though it were Jinglebob, Larn, and the other man he was yelling at.

I knew only that his wrist was working against mine, and it took all my strength to keep him from bearing down on Fuke.

Fuke had one hand on Roam's knife arm and the other clamped about the hand that held him down. I looked up and saw Trigueño standing in the doorway of the bedroom. He watched us passively, his hands invisible beneath his serape. War Bag was shouting something, but I couldn't hear. I wished Jack was with us, for he could have restrained both of them by himself.

Sutton entered the room, shoving past Trigueño. Before I could react, he had cast a bucket of lukewarm water over the three of us. Shocked, we all let go at the same time and fell back, spluttering and drenched.

"Y'all done made me waste muh warshin' water, now," Milt growled. "Roam Welty, you get yo black ass on down to the crik an' fetch me some more. Mistuh Fuke and Mistuh Stretch, I apologizes, but they just ain't no room for all this fightin' an' horsin' around an' shit amongst a couple a' old men like Mistuh Tyler and myself."

He looked over at War Bag, who was glaring at the three of us sternly.

"Ain't that right, suh?"

War Bag nodded, a thin smile barely visible beneath his whiskers.

"I had it in mind to talk to you boys about your shares," War Bag said lowly. "But you can't even keep quiet and listen a goddamn minute without takin' to the floor like a couple of kids in a schoolyard. I want Jack to hear what I got to say anyhow. You boys come back tomorrow."

"Hell with *that*, War Bag! I'm damn near *busted*," Fuke protested, his fight with Roam forgotten, and his hair hanging limply and dripping about him. "I won't have a roof over my *head* tonight without my *share*. There's a new *hunt* headin' out come New Years, and you're gonna be holed up in bed at *least* till then."

Fuke got to his feet and shook the wet hair from his brow.

"Now I appreciate ah . . . *comradery* as much as the next fellow, but my *livelihood* is at stake. We had an agreement, and now I *expect* to be paid."

War Bag's one eye stared at Fuke grimly. There was a look of grave disappointment on his face.

"I knew you were mercenary, George, but I thought you had more loyalty than that."

Fuke looked down quickly, then shook his head as though to cast off the guilt.

"*Money*," he growled.

"Awright," War Bag said, tired. "Awright. We'll make a new arrangement."

"You'll be laid up . . . ," Fuke began, rolling his eyes.

"Shut up! Just fasten your flappin' lip *one goddamned minute!*" War Bag exploded. It was startling to all of us and had the effect he wanted. Fuke's jaw clenched, and his eyes widened.

"I'll be bossin' a new hunt, soon as I'm well. Long before New Years, if I got anything to say about it. And I do. I'll take you all back on for full shares with a bonus for the one that gets the prize hide. But it ain't gonna be buff we'll be runnin.'"

We were all silent as the old man glowered. His whole body was trembling now, and his lowered brow heightened the menace in his eye. The light from the window played queer shadows across his face, crosshatching the jagged pink scar on his forehead. As we all watched, he tore off the kerchief, exposing the gaping, scabbed space that had once housed his eye, and thrust one shaking finger at it in indication.

"I'll pay double to whichever one of you greases the black faced red feather duster that done *this* to me."

His hand trembled, spilling the steaming broth all over his bedclothes, just as a doddering old invalid would. The stained sheets burned away the veracity that should have come with the passion that rippled through his voice.

He was utterly mad.

He stabbed one shaking finger at Roam.

"You know it now, Trooper. Black Face. He's been huntin' me all these years for what I did to him at Beecher's Island. I didn't believe it before. Not really. But after Still bit it....and...and my Billy....it was *him*. Black Face killed Billy too. I know it. I'm gonna take his scalp like he tried to take mine. I'll burn him down. *Burn that son of a bitch down to the ground!*"

He flung the bowl across the room and it clattered into a corner, bespattering the wall with yellow broth and strings of chicken meat.

He looked at us, his hair in disarray, obscuring the hole where his eye had been. His lips were flecked with spit and his chin whiskers dripped soup. His surviving eye was ablaze like a window lit by the bright light of madness in the brain behind.

"Trigueño's already signed on," he continued, spittle flitting from his quaking lips.

I looked back at the Mexican boy in the doorway. He stared solemnly at War Bag, unmoved by his behavior.

"Now...Trooper, I'm gonna need you to buy supplies. Powder. Get DuPont. It's dry. It burns best. And horses. George is gonna need a horse..."

Roam leaned forward.

"Sure, Eph. Sure. Jest tell me where the money's at."

War Bag grinned then, and it was frightening to see his expression change so suddenly. He shook his head like a child with a secret.

"No, you'll have to raise it, or buy on credit."

"What d'you mean?"

"I mean I can't pay you now. Not till we're out there. Not till it's done."

"What d'you mean? Where's the poor box?"

"I shucked it," he said, a note of triumph in his voice. He pointed out the window. "Out on the prairie. And only I know where."

Last Part: On The Prod

'Oh now we've crossed Pease River, and homeward we are bound.
No more in that hell-fired country shall ever we be found.
Go home to our wives and sweethearts, tell others not to go,
For God's forsook the buffalo range and the damned old buffalo.'
　　　　　　　　　　—from *The Buffalo Skinners*

First

"Why Do You Hate him so much?" I asked Fuke.

We were sitting on Milt Sutton's porch. It was night, and the land had gone dark and quiet. The lights in the Negro cabins were winking out, and the stars came out in their stead.

The light from inside Sutton's cabin washed over our backs and stretched our shadows out in the front walk. War Bag's news had convinced the old Negro to fish out a bottle of bitters which Fuke and I were sharing. Roam was cleaning the mess War Bag had made with water fetched from the creek. Trigueño was rocking inauspiciously in his chair in the corner of the room, as though waiting for something to happen.

"Y'know, the coloreds have a *saying*, down in the bottom lands," Fuke said. He was drunk now, and spoke with a voice that did not care whose ears it found. "'I'd rather be a nigger and work like *heck*, than a white trash cracker with a long red *neck*.'"

I blinked, unsure if he was going to go on. By now I was quite drunk myself, and my eyes were dancing along the faraway.

"I wasn't born in East Baton Rouge, as you might've *guessed* by now," Fuke continued, after a bit.

"Oh no?"

"*No*," he muttered, and paused, summoning the strength to say what he had to say. "I'm from *West* Baton Rouge," was what came out.

I blinked again, not sure of the significance, having never been to either end of Baton Rouge.

"My daddy was river *trash*. We all were. I *doubt* he could spell his whole name correctly. He did work no good *white* man wanted. He had hands like *shoe-soles* from pitching flour sacks and those blue cotton *bags* all day long down by the river. At night he came home drunk on *rotgut*. My family comes from a long line of trash, *Juniper*. My grandfather was shot by Sam Wilson for chasin' after a *nigra* woman at Cave in the Rock."

Fuke stared into the night. I was cautious not to interrupt, as though he were a rambling somnambulist. This was drunk talk—the most sacred kind, and not something to be questioned lightly.

"My daddy was a *good* enough man," Fuke went on. "He tried to do right by his *family*, and that was as much as you could ask for. A nigger roustabout ripped his belly open with a *gaff* when I was six years old, and his guts washed down the *river*. My family got split up and sent to *relatives*. I went to live with a cousin who was a Presbyterian *minister*. That's where I learned to read."

He looked at me for the first time then, and his light eyes, though watery and a little bloodshot, were more serious than I had ever seen them.

"But I'd ruther've *remained* an unlettered *ignoramus* and had my daddy and my little brother, and my *mama* all to me. I'd of forsaken all the goddamn letters in the *world* for that."

When it seemed to me he had finished, I ventured,

"Did you ever see any of your family again?"

"My mama was locked in a *sanitarium*. She never was in a state to receive visitors, you understand. Not that she would have *recognized* me had I come. I believe my brother *Henry* was sent to live with an aunt in St. Louis. I ran away from my cousin when I was *twelve*, but I never went looking for my brother."

I maintained a respectful silence before I decided to bring the conversation about.

"So, you hate Roam because he's a Negro, is that it?"

"Well," said Fuke, smiling a strange, amused smile, "he's a fair enough *scout*, but I don't care for his familiar manner. I believe it will serve *against* him some day. It was an *uppity* nigra like him that landed me in a stranger's house in East Baton *Rouge*."

His head lolled back lazily on his neck and he looked up at the sky for awhile and sang low, beneath his breath

> "I'm comin,' I'm comin,' though my head is hang-in' low,
> I hear the gen-tle voices cal-lin,' ole Black Joe...."

He stood up suddenly, and a little too quickly. He staggered off into the yard.

"Where're you going?" I called.

"Piss," he muttered, and soon there was the steady splashing of running water to prove it.

I listened to the stream of Fuke's release. I was flat broke. Not enough money to get me back to Chicago, not enough to move me on to a town of worth. I needed a job.

I looked over my shoulder. Roam stood in the light of the doorway, watching Fuke quietly.

As our eyes met, he turned away and went inside.

I spent the next day walking around Griffin trying to drum up employment. Sutton, who was a carpenter, was good enough to offer me a pallet on the floor of his cabin until such time as I could afford to stay elsewhere. It was a Christian gesture, but I told him I had to consider it. My few possessions were still at the Bison, and I wanted to see about a job.

As the day wore on however, Sutton's cabin seemed more and more appealing. No one had any work, and most confessed that they would be ashamed to pay me the wages they paid the many Negro laborers and odd-job Tonks to whom the menial tasks of water fetching, sweeping, wood gathering, and swamping the wagonyard fell. I had no particular skills worth offering above the level of labor, though. Griffin had no newspaper, and I had no other practical talents but what I had learned from my compatriots.

I considered Jim White's offer, but December was yet a long way off. Passing by the Bee Hive, I found Fuke engaged in his new profession at the faro table. He seemed content and was making cash. Fuke had not spent the night at Sutton's, but had wandered off drunk all the way back to town.

I went to Shaugnassey's and found Jack awake and supping on a stew Aunt Hank, the lady who ran the Occidental Hotel, had brought him. He had a big goose down pillow and a homemade comforter on his cot now. He was bandaged and his wounds were cleaned, but he was dismayed at his incarceration.

"If'n that damn dawg hadn't 'a kept on comin' at me after Ah got Whisper outta his jaws, Ah shore wouldn't be in this fix. Whar is Whisper anyhow?"

"He was under the boards in front of the Bee Hive yesterday. He gave me this," I said, displaying my slashed knuckles, "when I tried to get him for you."

"Reckin Ah'm right sorry, Stretch."

"It's alright. Jinglebob's sure not feeling kindly towards you, I'll tell you that."

Jack rubbed his bandaged leg thoughtfully.

"If Ah had it do all over again, Ah'd've whupped *him* upside a post." After a moment though, he rescinded some of his malice. "Ah do feel bad about his dawg, though."

I told him about War Bag and his condition, and about our lost money.

"Ye really thank he's crazy?"

"Most definitely, Jack. He says he wants to go out and hunt the black faced Indian. He tossed the poor box away, somewhere out on the plain."

"Ye shore them Texans didn't smouch it?"

"He admitted it himself. He did it so we'd have no choice but to follow him. Said it was his plan."

"Right clever for a man out of his head," Jack mumbled.

"Not hardly. He was delirious when he did it. How could he know where he left it, or how to find it, or even if it would be there after so long? It's crazy!"

"Well, he has an eye for the land, Ah reckin. There ain't so many folks out lookin' for a lock box out in the middle of nowhere, is there?"

"Aw, you're just as crazy as he is, Jack."

Jack nodded to himself.

"Ye seen that black faced Injun, didn't ye?"

The image of the Indian's pitch face as he cast his lance down into Frenchy appeared instantly in my mind, and would not be denied. By my hesitancy, Jack had his answer.

"Ah thought ye did. Ah reckin we owe it to him."

"Who do you mean, Jack? War Bag?"

"Well, him too, but Ah was talkin' about the Injun. We owe him. For Monday. And that French feller too, Ah guess. Is there any walnut trees growin' roundabouts?"

I blinked.

"What?"

"My grandmaw used t'say walnuts was brain food, for them that was teched. Ah don't know fer shore, but if'n ye crack 'em open, they shore do *look* like brains. Hey Stretch... ye been sleepin' awright?"

Jack's train of thought was sometimes hard to catch.

"Why do you ask that?"

"Well, Ah been having bad dreams. About snakes...."

Just then Dick Shaugnassey ducked his head in behind the curtain.

"I just spoke to Mr. Dowell. Those colored boys are back with the stone for the new jail. I'd like to be able to get at my liquor without stepping around this big ox, so I've convinced Cade Reddick to let him pitch a hand in buildin' the jailhouse. Big strong fellow like you ought to have it up in no time at all."

Jack sat up, the light of excitement in his eyes.

"Well, Ah'd shore appreciate the work. Cain't abide layin' around all day on account of a pinprick and a couple o'dawg bites."

"There's a good lad," Shaugnassey said. "The Deputy will be around about noon to fetch you, and he'll have you back by evening."

"Ah'll watch for him," Jack nodded.

Shaugnassey ducked back out.

"Looks like Ah got a job already," Jack said to me in all seriousness, rubbing his hands together.

I went out to have a word with the jailer.

"Mr. Shaugnassey, is Jack going to be recompensed?"

Shaugnassey just cackled as he stepped behind the bar.

"*Recompensed!* Well now, that is a fine, expensive word. It's Juniper, isn't it? That's what your friend calls you? It suits you."

There was a dragging sound, and the curtain opened behind me. Jack had pulled the cot to the doorway to speak.

"Stretch, would ye try to fetch Whisper for me agin?"

A week went by, and I neither caught Whisper nor found employment. I continued to stay in the Bison. The money I had got from Roam steadily dwindled. I purchased a bar of lye and bathed in the Clear Fork, and the ink of a hundred filthy days clouded the water around me. I scrubbed my clothes and beat them dry on the rocks, feeling refreshed and invigorated. I thought my rise in cleanliness might endear me to a would-be employer. It did not.

I do not know where Fuke spent his nights, but from the seedy company he grew to keep, I can guess. He and Pat Garrett were as chummy as ever, as was Frank Collinson, the Yorkshire Englishman who had invited him to participate in the December hunt. Fuke took to calling him 'Yorkie.' Cathy was almost always hanging about his neck and laughing at some witty utterance of his.

Despite his rather personal revelation to me, I made a point of avoiding him. It was childish really, as he had no idea I'd been smitten with that particular dance girl and she obviously had no memory of the best dance partner she had ever had. I wondered if that had been a line she fed all her dancing partners. I'd thought I was a decent dancer. I didn't set foot in the Bee Hive, but I was often in Shaugnassey's, asking about jobs and visiting with Jack most nights.

Reddick said Jinglebob had decided to press charges formally against Jack, but I did not notice the albino about town much. John Larn and their crazy eyed friend, who I learned was John Selman, I saw only rarely.

Jack and the two Negro laborers, Hank and Sway, undertook the construction of the jailhouse. Ironically enough, Jack would be its first inmate. Sheriff Jacobs and the posse showed no signs of returning, and the circuit court was not scheduled to pass through Griffin for another week. I saw Milt Sutton now and again on my visits to the construction site, as he was cutting a door for the jail.

He told me that while War Bag seemed to grow stronger daily, he spoke of nothing but preparing for his "hunt." He was always whispering to the boy Trigueño in Spanish that neither Sutton or Roam understood well enough to follow. The old man sent the boy on mysterious errands all over town. I saw him now and then at the wagonyards, at Conrad's, and oftentimes going back and forth from the Tonk village at the base of the Hill with some parcel or another under his arm.

Roam did his best to continue caring for the man.

"Why does he do it, Milt?" I asked. "For that matter, why do you let him stay there?"

"Aw," Sutton would say, "well...Roam Welty done a good turn to one o'my kin oncet, an' I reckin Mistuh Tyler done right by him. That's good enough for me."

"What sort of good turn, if you don't mind me asking?"

"Naw, suh," he said, taking off his stiff brimmed hat to run his hand over his scalp, "I don't mind none a'tall. My nephew Mervin, he was a soldier with Roam in Kansas."

"In the tenth cavalry?"

"Naw, this was before that. In what was called the Indian Home Guard, during the War. Mervin, he was slave to a Seminole family over in Indian Territory..."

"An Indian family?" I interrupted.

"Yassuh," he confirmed. "When the Secesh come to the Creek Nation and asked 'em to join up and fight for the Rebs, they was a chief called Opothlayahola that took a passel of Indians and some colored folks and headed up North to join the Union. That's where Roam and Mervin met, I reckin. They was chased by Confed'rate cavalry the whole way, and when them that was left finally made it to Kansas, well, that's how the Home Guard come about. They was jest boys, then. Roam took a Rebel ball in the leg for my nephew."

"So he saved his life?"

"Oh yassuh," Sutton confirmed. "Mervin told Roam he was welcome to come on down t'my place any time he wanted. That's how I met him."

I was going to ask about Mervin, but Sutton interrupted. "Mistuh Stretch, now how's 'bout you tell me somethin'?"

"What's that?"

"That Mistuh Fuke. Whar he say he come from?"

"He says he's from East Baton Rogue," I said, a little confused. "Why do you ask?"

"Aw," he said, waving away my question and whatever curiosity he had. "Nothin.' He jest talks real nice, s'all. Now tell me somethin' else, suh. When you goin' come an' stay at my house?"

Truth tell, I had no more compunctions against taking him up on the offer, except that I feared War Bag and what he might do in his state. Maybe sit out on the porch with Bullthrower and pick off anyone that came close, while that half-wild Mexican boy kept him loaded and watered his barrel. I considered telling Deputy Reddick about War Bag, and perhaps reserving him an accommodation in the new jailhouse. I just smiled and told Sutton I would see him again later.

At the end of the week I told Windy Bill I would join the Mooar outfit in December if they would have me, and that I would be staying at the Negro carpenter's cabin in Darkytown. I told Jack I was moving out of the Bison and would take his things out to Sutton's for safekeeping. I made one more attempt to catch Whisper.

Dusk came up purple that evening, and I stopped at Collins Creek to wash the fresh blood off my poor hand. From there I hoofed it to Darkytown, bent under the weight of both mine and Jack's gear. The sun had gone west by the time I was winding through the black cabins, and it was very dark when I came to within sight of Sutton's. The light in the bedroom window was lit, leaking a pale glow through the cracks of the shutters. War Bag was awake inside, plotting, or else he had fallen asleep with the lantern lit. The sitting room was in darkness, as was the tool shed where Trigueño slept. His horse was still there. There was no moon.

I could see the orange glow of two cigarette ends on the porch. They sank to knee level as the smokers sat down on the porch and began to talk. I could make out Roam's voice, and knew him immediately, but the other man was not Sutton or War Bag.

Then I stopped in my tracks. Tied to the front fence was Othello, my stolen horse.

Second

"Dobie," Roam said in the dark. "Been a long time."

I crouched down where I was, and let my bundles slide noiselessly to the road. The other man's reply was lost to me as I crept along on my haunches till I was laying in the tall grass that gathered around the post oak kitty corner from Sutton's front yard.

I could see clearly that the horse out front was indeed Othello, my own mount that had been stolen in the Indian attack. There was my saddle on his back, and my blanket rolled behind the cantle. There was a buckskin rifle scabbard hanging there that was not mine, and the stock of the Spencer inside was decorated with brass tacks.

"I seent your brother," Roam said. "At the Walls."

"Them fuckin' crackers burnt him down," affirmed Dobie, the other man. "Told him not to get so close. Told 'em all."

"I seent he was buried," Roam said. "Didn't let them do nothing to him."

"Thanks for that," said Dobie.

I peered through the space between the fence rails, and tried to will myself to see through the dark. I could barely make out the form of another Negro, lighter of hue and similarly dressed to Roam.

"What was you doin' there?" Roam asked.

"Jest tryin' to stop it fore it started, I reckin. You know how it'll be once the white folks start to pourin' in, they won't stop killin.' And when the buff gone, we gone."

"Never thought I'd see you again," Roam said.

"Well," Dobie answered. "Shore never thought I'd be sittin' on no porch talkin' with you neither."

"Why you down here, Doe? Why the Cheyenne down here? And don't tell me 'bout no buffalo."

"The war chief. He a old time Dog Soldier. He led us down here. Meaner than hell."

"What war chief?"

"His name Black Face."

I felt my veins ice over.

"He goin' crazy. All the nations goin' crazy. It's them Comanche. They got the other tribes all fired up. This medicine man, Isatai, he whupped the Kwahadi into a lather, and they talked the Cheyenne into it. Arlo and me was runnin' with Graybeard's bunch. They was Dog Soldiers mostly, and Black Face was one of 'em. We was supposed to burn the Walls down, drive them hunters back up to Kansas, or kill 'em if we could. But it went all wrong."

"I seen that," Roam agreed.

"Goddamn Isatai, he got 'em thinkin' he could stop bullets with paint." Dobie continued. "And Arlo, well . . . . you know him and his fool ideas. He reckoned he could train the Dog Soldiers to answer to the bugle and fight like white soldiers. But it was like them hunters knowed what was comin'."

I remembered the Irishman at Adobe Walls, and his story about the sound of the bugle preceding the charges of the Indians.

"I told him not to get close. But we'd just took a load of coffee beans offa this buffalo outfit, and that stupid nigga wanted sugar."

"What about Black Face?" Roam asked.

"After the Walls, we picked up your trail. Greybeard din't want no more part of them Sharpes rifles. Black Face, he argued with him. Said he hadn't got no scalps on the whole trip, and only had empty ponies to show for it. So he took his men and split off, and I went with 'em. We went south, and struck your trail. 'Didn't know it was you. One of your boys split off, and Black Face sent me after him. I chased him down, but my hoss thew me in a prairie dog town."

I crept in silence along the rail, closer to Othello and the front gate. My hand closed around the handle of my Volcanic.

"I didn't get there for the fight, but I seen you when they was bringin' the hosses back. Black Face, he was crazy mad. Somethin' about a white man he knowed . . ."

"Who?" Roam whispered, and I could hear the cold creep into his voice. "What white man?"

"Long time ago Black Face got shot in the side of the mouth at Beecher's Island. That's how he got his name. He still got the mark. He say the old man you with is the one done it to him."

I had my pistol halfway out of my holster when Roam stood up off the porch and began to pace. Othello nickered apprehensively.

"He got us camped jest a few miles outta town," said Dobie. "They's goddamn cavalry all over the place day and night. But he say he followin' a vision. He say he goin' kill that old man."

Dobie cast his dead cigarette beneath his foot and ground it out.

"He want us to burn the town. He say his dream tell him his oldest enemy is goin' die by fire."

"How many braves y'all got?" Roam asked.

"Ten. But Black Face sent two of 'em to parley with a band of Comanche we spotted two days ago. They's about thirty of 'em. Black Face goin' play up the hate they got for the Tonks, try to get 'em to join us. Some of the Dog Soldiers lost kin fightin' your outfit. They want blood, Roam. They goin' do it."

"We'll tell Major Schwann up at the post. Where they at?"

There was a pause, and I was trembling and perspiring, though the night was cool. If I cocked my gun I would be heard.

"Naw," said Dobie.

"What?" said Roam.

"I ain't come here to turn 'em in, Roam. I come to tell you you got to get outta town. You cain't tell them soldiers."

"The hell I cain't."

There was a short scuffle, as Dobie stood up and grabbed a hold of Roam's arm. I rose up to my knees and balanced my pistol on the fence rail, trying to let the light into my eyes.

"They my tribe, Roam. They my brothers. I cain't turn my back on 'em. 'Sides ...Arlo was my lil' brother. He was all the family I had left, and he gone. We goin' burn this town down to the ground. You can b'lieve that."

Roam shrugged his arm out of Dobie's grip.

"You crazy too!"

"Naw, nigga, *you* the crazy one. Been runnin' like a trained chimp for the white man since you lit out the Nations. Ain't you learned nothin? Ain't you seen nothin? They settin' us agin each other like dawgs. Jest like two dawgs—red and black! We ain't goin' win nothin' s'long as we killin' Indians instead o' crackers. Out there, out there with my brothers....I ain't no dawg. I ain't no dawg and I damn shore ain't no nigga."

"They's a passel o' white folks," Roam said. "Soldiers, needle guns . . . .you ain't goin' burn these folks out. They goin' be here till kingdom come."

Dobie backed away from Roam, and I could see them staring at each other in the gloom.

"Maybe so, Roam," Dobie said then. "But they goin' bleed for it."

Dobie turned to walk away then, and I stood up from my position and aimed my Volcanic for the center of his broad chest.

Dobie looked at me, and I saw his white eyes loom. I felt my thumb tug back the hammer, heard the mechanism click, felt the pistol awake and alive in my hands, ready to strike him dead on my command.

In the time it took my mind to register that he did not care that I was going to kill him, he rushed me. His eyes told me he was going to kill me, and no amount of lead I could pour into him would stop him.

He took me by the throat and jerked me over the top rail of the fence. The bones of his Indian breastplate rattled musically as he moved. As I hit the ground the Volcanic slipped from my sweaty hands. Dobie bent over me, an immense figure. Elk teeth and porcupine quills dangled before my nose like a child's mobile.

"Goin' wring yo neck, you *skinny white chicken*," he rasped, strangled by his own immutable hatred.

He had a knife. He could have cut my throat or slashed the gun from my fingers. But he had dragged me down and now sought to squeeze the breath from me.

As I writhed beneath him, all pretense fell away. All the self-delusion I had ever spewed. All the rhetoric I had taken from greater minds and adopted as my own faded from my selfish heart like meaningless words copied in a childish hand with novelty ink.

I was struck down in the basest moment of naked fear and outrage. Outrage, not that I would die, but in my astoundingly Narcissistic brain that I would die unknown and unpublished (!) in the yard of a carptenter's cabin in the Darkytown of a backwater settlement at the hands of some renegade nigger. I was destined to yet fall lower; so low that I cannot look back on that moment now without reliving the intense and crippling shame which I felt then.

In that reprehensible dungeon of thought, I was laid low by Roam Welty. Roam, to whose aide I had hesitated to come twice already. He stepped forward with his pistol drawn and said, "Leave offa him, Doe."

My face was purpling, but even the pounding blood in my ears could not block out that cool decree.

"Dobie! Leave offa him or I'm goin' put you out!"

The thumbs boring into my throat relaxed, and air hissed into my ragged throat hot as whiskey. My tongue curled in my mouth, thick and gagging. With that final disgusted push of a man finished with dirty business, Dobie cast me down and stood up.

I coughed, the blood draining dizzily from my beating

temples. I curled on the ground, my throat aching and bruised.

"Get outta here, Doe," said Roam with a flick of his pistol.

I heard the saddle creak as Dobie mounted Othello.

"Send that old man out, Roam," he said in parting. "Be the only way."

I heard the clop of hooves as he went off down the road into the dark.

Roam knelt beside me, and his hands touched my shoulders.

"You awright?"

I had no words for Roam just then. I thought of all the times I had neglected to stand beside him, and of the reason why. It was not just cowardice. Down beneath the idyllic surface of what I had built to be my progressive and transcendentalist heart had lurked the plain truth that I did not deem a black man's life worth my own.

I had treated Roam with a disrespect more vile than even Fuke's, for it was basted in self-delusional sweetness. Despite all he had taught me and done for me, and though he had never given me reason, I had still suspected him of stealing the poor box. I knew in my heart that although I had always admired Roam, I had resented and feared him too. Though my father had bought off his prejudice with abolitionist contributions, he had still impressed upon me the lesson that a Negro was something less than a man. He was a charity. Something to pity and support, like a dwindling colony of impressive wild animals which ought to be allowed to live 'somewhere,' so long as it wasn't our own backyard.

Roam had just saved my life. It had been a dilemma that in his place, I would have protracted to the length of Hamlet's. But for Roam it was simple. Like the mad old man in the cabin, and even Fuke, who had nothing but contempt for him, for all my faults I was still his partner.

Roam hoisted me up and drew my arm over his shoulder. I felt as though I had misplaced the lower half of my body.

"It's all true—What War Bag said. Isn't it?" I whispered, as he helped me to the porch.

"I 'spect it must be," he said, after a moment.

He propped me against the post.

"Set here."

Roam returned with a pan of water and a rag, which he wrapped around my neck. Sutton was with him, sleepy eyed and bearing a lantern. The light under the door of the bedroom shined where War Bag nursed his wrath.

"What's happenin'? Thought I heard somebody just a while ago."

"Nothin,'" Roam said.

Roam shot me a look that told me I should keep quiet about Dobie Flood and the war party. I decided he knew best.

"Nothin' oughta be quieter'n that," Sutton hissed. "I cain't hardly git no sleep with all this nothin' goin' on."

"Sorry, Milt. We be beddin' down soon," Roam said.

"What's a'matter with Mistuh Stretch there?" he said, leaning in to peer at me.

"Whiskey," I said, forcing a smile, though my throat was in agony.

Sutton broke into a gummy grin and laughed.

"Yeah, she'll do it sometimes. Well," he said, straightening up, his bones popping. "G'night then."

"Night, Milt," Roam said.

"Good night," I rasped.

Sutton disappeared into the house. I leaned in close to Roam, my voice low.

"What'll we do?"

"In the mornin,' you get on up to the post and tell Major Schwann."

"What about you?"

"They won't listen if I tell 'em," he said. "I'll go with you, though. They might not be no trouble if Black Face don't get them Comanches on his side."

"When do you think they'll come?"

"Not tomorrow. It'll take a whole day of parley to get them Comanches goin.' Then, maybe on Monday mornin,' if they really goin' do it. Yeah. It'd be Monday mornin.'"

Sleep was long coming that night. Each time I managed to close my eyes and felt my self slipping away, the notion came to me that this was what death would be like. I kept seeing Monday's eyes as the cognizance faded from them like a glass obscured with a puff of warm breath. If I were to die, I decided I wanted to die quickly, as Frenchy had, blasted into oblivion with no time to linger on my life and the fear of what might lie beyond. I drifted off several times that night, only to jerk awake.

Morning seemed to come too soon, and the cool night wind settling the cabin and whistling over me from under the door did not soothe. The room was a blue haze when the toe of a boot prodded me awake.

I looked up, ready to bid good morning to Roam.

Standing over me, garbed in his corduroy coat and bearing his Henry rifle in the nook of his arm was War Bag.

Third

"Get up," War Bag commanded.

His stringy white hair hung down like a raged curtain framing his solemn face, monolithic and grim in the early light. His one eye looked down on me.

I glanced across the room. Roam was still asleep, lying on his side facing the wall. Somewhere Sutton's snores came to me from the next room. It would only take a peep from me to get Roam on his feet. I looked back up at War Bag.

The old man had read my intent, and shook his head no.

I slowly pushed back my covers and got to my feet. My gunbelt hung on a chair in the corner. My eyes went momentarily to it.

"Take it," he said. "And come on."

He went outside.

I took up my pistol belt with its empty knife scabbard, and turned again to look at Roam on the floor. I need only say his name to wake him. But something in the resonance of the old man's voice... It had returned to its former greatness. No more was it a frail, convalescent whisper. But he was mad, I told myself.

I went outside, buckling on my pistol as I went.

Trigueño waited in the yard, holding Solomon and his own horse by their bridles. He glanced at me and at the old man questioningly.

"*Está bien*," War Bag rumbled.

I stood on the porch as the old man took the black horse by the bit and led him out into the road. His hat was in his hand, and he placed the battered thing squarely on his head and turned to me, the pre-dawn shadows hiding his brow beneath the nicked and battered brim.

I stepped off the porch and went to him. Before I was there, he had turned and was walking up the road. Trigueño led his horse alongside.

Their saddlebags bulged with supplies, and their bedrolls were tied behind the cantles. Stretching out like a great flimsy sliver from a loop on Trigueño's saddle was his *cibolero* lance, quivering with every step of the horse.

We walked on in silence between the quiet cabins as the Sunday morning sun peeked over the eastern horizon, a faraway beacon of bright orange light in a deep blue sea.

Some of the Negroes who worked cattle were saddling up to ride out to their bosses' ranges. The smell of bacon frying snaked through my nose. Otherwise, the world was asleep.

We crossed Collins Creek as if in a dream. I did not feel any fear, but a weird otherworldly inevitability. I knew I should run back to Roam, but I wanted to see what War Bag was doing, and I thought it prudent to remain quiet. As it had always been, I was at his call.

We proceeded on through the business district, which seemed ghostly without the jangle of the tracer chains and the rattle of the wagons. The horses stirred in the wagonyard as we passed. The wet earth of the street sucked the hooves of Solomon and Trigueño's mount.

Down the avenue there came a strange, metallic rhythm that desisted at regular intervals, only to start again. As we passed the Bee Hive, I saw it was the bear. He was awake and rattling his cage. He stopped as we passed, sniffing the air, then resumed when we showed no interest in him. The sound faded as we continued on.

Up on the Hill, the bugler blew reveille. I watched the flag hoisted up the pole, its red, white and blue muted against the lightening sky. Rather than take the road that led up to the fort proper, we veered into the sleepy Tonk village.

From what I had seen of the Tonkawas I was not inclined to regard them as early risers. I was surprised to see one of the old buffalo hide tepees glowing from within. The tepee was in itself a unique sight, for most of the other domiciles were crude thatch or hide wigwams and sod huts covered with dried grass and timber. They resembled the brush burrows of field mice and were about as clean.

This tepee rose like a regal lord, and the effect was furthered by the firelight and the majestic wisp of rolling smoke that rose from the top, falling out over the morning sky and fading away. There was a queer, rhythmic beating emanating from inside, and we could hear low musical voices as we approached. It was this that War Bag had brought us to. There was no doubt.

The shadows of the few figures within danced across the translucent surface of the lodge, like mystic images rendered in magic paint. One of the figures stood and emerged, offering a brief glimpse of a swaying flame and huddled, silver haired men in buckskin.

The Indian was a man of War Bag's age, proud in his bearing and white-haired. He wore a cowl made from the eyeless head of a wolf, and the pointed ears stood sharp against the glow of the fire. The tracings of a beard hung like cobwebs from his chin, and his lined face was bisected lengthwise by a series of intricate black tattoos. He saluted War Bag after a fashion, and the two of them conversed in hand signs.

I glanced nervously at Trigueño. My scalp prickled with a supernatural dread.

The old Tonk gestured then to the lodge, and two more Indians emerged. These were thin, grey haired old men. Their bony wrists and faces were tattooed with savage patterns similar to the head man's. I started, for they were armed to the teeth, their tightly cinched belts sagging with knives and cap and ball pistols. Poorly garbed in mismatched clothing, their costumes attested to some past military service. One wore a blue felt hat over a pale yellow cavalryman's kerchief tied in a rude band around his head. The other sported a worn-out voluminous Union blue great coat. Their belts were U.S. issue, the letters on the tarnished buckles partially scratched away in some deliberate act of vindictiveness.

The black eyes of the scraggly pair passed from War Bag to Trigueño and flashed like animal teeth at me. My spirit shriveled before those savages, as they had underneath Dobie's naked hate.

The old Tonkawa spoke then, and placed his hands on the two mens' shoulders, intoning in his own language, apparently to impart some benediction.

A woman draped in a blanket appeared from around the end of a lodge, leading two shaggy ponies with rope bridles and hemp reins, patterned blankets thrown over their bowed backs. The two old men took the horses from the woman, who then expeditiously departed.

War Bag passed his hand one over the other, and the old Tonk repeated the gesture and nodded. Without another word, he ducked back into the tepee and let the flap fall shut behind him.

War Bag nodded once to the two old men, then turned about face and walked past me out of the village, leading his horse. Trigueño followed, and the two dour-faced Tonks went after.

I hurried to catch up to War Bag.

"What's going on?" I asked him as we left the village.

"Just signin' on a couple of hands," War Bag said.

"Those? What can they do?"

"They're hunters," War Bag said.

"Last night you heard Roam and Dobie talking."

War Bag said nothing.

"You know what'll happen. We've got to warn the post."

"Go ahead," the old man said.

"I will!" I declared. I broke off from the macabre little procession and hurried up the hill road.

"Stretch!" War Bag's voice boomed.

I halted and looked over my shoulder. He and his little army had stopped, and were looking at me.

"Where's Fat Jack?" he asked.

"They have him locked up at Shaugnassey's," I said.

War Bag turned his head to regard the Flat. A little boy came marching up the trail from the Tonk village, yawning and rubbing his eyes. He had a switch in his hand. The same he had been using to poke at the caged bear with all week. He skipped past us without a glance and scurried towards town.

War Bag watched the boy, and then followed in his steps. Obediently, his retainers fell in behind.

I renewed my hike up the hill.

The post was strangely quiet so soon after reveille. One or two soldiers on water detail bore empty buckets down to the river. They ignored me as I rushed past the dim locked store and clomped into the adjutant's office.

I found a surly major sitting at his desk, peering thoughtfully through his wire rimmed spectacles at the individual components of his disassembled revolver, laid out in the tidy manner of a watchmaker upon a white handkerchief. No one else was about. The major looked up as I came in, out of breath from the brisk walk up the hill.

"Are you Major Schwann?"

"I am," the major replied. He was a Yankee like me. "What can I do for you?"

"You've got to mobilize the men," I implored.

Schwann sighed and sat back in his desk.

"Would that I had the authority to do so, sir. But you see, Colonel Buell, who is in command, being a Colonel, is not subject to obey orders from myself." He fingered his shoulder epaulets and adopted a slow, patronizing tone, apparently for my sole advantage. "I am a *major*. In the military, we have what is known as a *chain of command*. It is for the benefit of *officers*, so that we may deduce to whom among our brotherhood it is that we are bound by regulation to salute, and to whom among the enlisted men it is our privilege to demand of that they salute us."

Vexed, I leaned on the desk to better emphasize my purpose. Major Schwann recoiled slightly, and glared down at my offending hands, which had inadvertently upset the corner of his handkerchief and the springs and rods thereon.

"Major, I'm aware of your rank," I said, trying to put

across that I was not the dim-witted herder he seemed to think I was. "Where is Colonel Buell? Where are all the troops?"

"Not that its any business of yours, sir," Schwann said, his tone as even and unhurried as before, "but Colonel Buell has taken six troops of our cavalry divisions and two of the infantry companies out on maneuvers."

"What does that leave you with?" I asked, feeling a sinking sensation.

"Work details," he said, rearranging his pistol pieces and flattening out his handkerchief. My hand, still on his desk, blockaded his efforts. He turned his eyes up to me expectantly.

I straightened up and put my hands at my sides.

He returned to his work without a word of thanks.

"How far out are they?" I pressed. "Can they be recalled?"

"I would not be at liberty to divulge Colonel Buell's whereabouts," he said, polishing his cylinder and not looking up at me, "even if I were privy to them. General Sheridan is mounting an offensive against the Comanche confederation in which Colonel Buell's command is no doubt playing a decisive part. Meanwhile its is my own duty to assure that the stench of the compost heap is conspicuously lessened in time for his return."

"Can't you spare any men for the field? Or to mount a defense of the town?"

"The town of Griffin falls under the jurisdiction of the Marshal. His own title gives him that distinction. Having a military title, I am only bound to concern myself with the Fort of Griffin. Why are you pestering me with all this, sir?"

I had thought it would never even occur to him to ask.

"There's a party of Cheyennes whose intent it is to fire the town."

Major Schwann's eyes flashed, but cooled almost immediately.

"I see. And have they announced this intention publicly, or have they only told you?"

"I just know they are."

"Ah."

"They're Dog Soldiers. Maybe a dozen. Even this minute they're parleying with a band of Comanches. Thirty or more. They're going to attack the town Monday morning."

"How did you come by this information, sir? You don't look like a scout."

"It would take too long to explain. How will it look if the Flat is burned underneath your nose when you had ample warning to mount a defense?"

"Some in the War Department might throw a gala."

"How can you say that? There are people living down there!"

I slammed my fist down hard on the table, causing all his pistol parts to leap and jumble together.

"See here, sir!" Schwann erupted, jumping to his feet. It was the most animate I'd seen him yet. "Remove yourself from this office before I have you removed forcibly!"

"I'm telling *you*, sir. Those Indians are coming, and all you're intent on doing about it is to keep sitting here fiddling with your pistol!"

"Firstly, no Dog Soldiers have been seen this deep in Texas *ever*! Second, there are more soldiers than Indians in Texas right now! To think a party of savages would risk the trek into West Texas just to fire that insignificant little den of niggers and guttersnipes is preposterous! Now if you please, return to the Bee Hive and inform all your "barroom buddies" that their shenanigans will not be tolerated by myself or my administration simply to satisfy their inane wagers. They would probably find their money better spent consorting with the crib girls, who are *this* close (he held up his finger and thumb) from being ejected from the land for contaminating our troops!"

Furious, I turned and stormed out.

I stumbled down the hill in haste. War Bag and his cadre were nowhere in sight. I thought of whose assistance I might enlist. Deputy Reddick and Mr. Stribling had the ranking authority to order the locals into a defensive posture, but there was no guarantee they would be obeyed, and even less that Reddick would be capable of organizing anyone. The vigilantes were the best bet, I decided. Though I only knew Larn, Selman, and Jinglebob Beddoe, I had seen some of the others and knew them by face. If at all possible, I would approach them first, for I did not relish coming to men like John Larn for aide.

When I reached Griffin Avenue, it was just in time to see John Larn and John Selman at the head of a posse of twelve riders. They galloped off in the direction of the river crossing. To my surprise, I spied Deputy Reddick among them.

Windy Bill was standing in the street, waving his hat to them as they went.

"Make us proud, boys!" he called. He was already drunk.

"Windy!" I called, rushing over. "Where are they going?"

"Aw, they got a line on somebody runnin' off a remuda up at the Treadwell Ranch."

"Listen to me, somebody's got to catch them and bring them back."

"Slim chance of that, amigo. They got the smell of blood. Even Reddick. Blossoms'll be bloomin' down by the river."

"Windy, shut up for a minute," I said, grabbing hold of him to get his attention. "Listen to me, I want you to gather all the hunters together at the Bee Hive."

"Shit, Stretch. Most of 'em are there already..."

"Get everybody from the eatery and the hotels too. We gotta make plans!"

I rushed off down the road, leaving Windy Bill in a daze.

"Plans for what?" he called.

But I was running. The Tonk boy was crying in the street, carrying the broken halves of his switch in each fist.

The sun was up by the time I was running down the road through Darkytown. It was an overcast day, very cool, and the sun could not penetrate the gray clouds overhead.

I came to Sutton's cabin. He and Elijah, the Negro barber, were loading the finished door for the jailhouse onto a wagon.

"Milt, where's Roam?" I called, as I fell against the fence, winded.

"He left not long after sunup. 'Said he was goin' lookin' for you," Milt said, scratching his head. "Mistuh Tyler and that Mex boy is gone too."

I cursed and turned to run back, but paused, remembering Jack's Winchester and Frenchy's pistol. I rushed into the cabin and got the belt from which hung the pistol and Jack's Indian knife, and took the rifle and the saddlebags, which had the cartridges and his skinning kit.

I raced back out the front walk and onto the road.

"If'n you wait, we give you a ride!" Sutton called.

But I was gone.

I remembered the race with Caddo Charlie. This time my opponent was not one Indian, but anywhere from ten to forty, and it was lives, not money in the balance. The run was harder and longer, and I was burdened with iron and leather. My feet were heavy as stones and there were cedar logs under the skin back of my legs when I reached Collins Creek for the third time that day. I stopped to drink, and watched the clouds curling overhead in the reflection of the water. A black boy looked up from his fishing, startled to see me slurping from the creek.

"You oughtn't drank outta there, mistuh," he offered. "It's cleaner up the crik awhile."

I thanked him and headed on, shouldering the rifle.

I was pleased to see a crowd gathered in a semicircle out in front of the Bee Hive. It seemed like all the men that remained in town were gathered. I blessed Windy Bill's name and rushed down the street.

Everyone was talking at once, and took no notice of me as I came up.

"Alright everybody...!" I began, even before I had pushed through the crowd. Several men turned around and shushed me.

"Hush up, boy!"

"Yeah, button up awhile, we can't hear as it is!"

I skidded to a stop, perplexed. It was not me they were awaiting. In the center of the gathering, Jack's head stood above the crowd, his eyes intent on something else sharing the circle with him. Someone was speaking, but I had to strain to listen above the murmur of the crowd.

"Alright, alright, we'll be ready to begin in just a few moments. Now then, which of you men thinks their knife fitting to match against the claws of this beast?"

Of the thirty or so men on the street, nearly three fourths reached down and pulled out their Bowie knives and held them aloft. Everyone was talking and laughing at once, and the steel points gleamed against the sky. It was like witnessing a ritual of some bizarre Hindoo murder cult.

I saw Jack's head turn to face the crowd, his eyes moving among the proffered knives.

Then I heard the bear groan.

Fourth

The Bear pawed at the iron door of its cage, curled claws raking along the bars. Of the two Jayhawkers who had caught it, one stood grinning like an imbecile, his hand poised to slip the bolt on the cage door, while the other worked the crowd.

"Go on, Mr. McDade," he said to Jack, who was stripped to the waist. "Pick a winner."

Jack was watching the offered knives with selective interest when I burst roughly through the gathered bodies and stumbled into the circle.

"Curb your enthusiasm, sir!" the Jayhawker said, smiling at me.

I looked past him.

"Jack!"

"Hey Stretch, is that my knife ye got there?" Jack said, pointing to his belt, which was draped over my shoulder.

"What the hell are you doing, Jack?"

The Jayhawker stepped between us.

"Now, now, Mr. McDade has agreed to this little contest of his own free will, isn't that right, boys?"

I was assaulted from every side with affirmatives, peppered with demands that I 'get the hell out of there' and 'quit holdin' up the show.'

I stumbled against Jack, seizing his bandaged arm.

"Jack, you're not going *through* with this?"

Jack looked from me to the bear. They had starved him in that cage for four days, and poked him with a stick. He would rive and devour the first thing that got in front of him.

"He'll kill you, Jack," I said, pulling his head down to me so I could whisper. "You've got to tell them you were just fooling."

"Don't reckin Ah can, Stretch," Jack confided to me in a comparable whisper.

"Why?" I asked, gripping his shoulders, suddenly enraged. "Jack, what'd they tell you? Did they tell you they'd let you go if you did it?"

"Naw," Jack said, furrowing his ponderous brow. "Ah got to wait for the judge."

"Whose idea was this, Jack? Who put you up to this?"

"Nobody," Jack said.

Behind me, I felt many hands grip my limbs.

"Cain't ye see? Thet's *my* bar. Ah cain't let 'em treat 'im no more like they done."

I was dragged out of the circle. Hoisted high over the crowd, I was passed from hand to hand and then roughly deposited at the back. I dashed the back of my head on the lip of the boardwalk and curled up in the gutter, clutching my singing scalp. Blood ran between my fingers.

I sat up slowly. The belt was no longer around my shoulder. I scrambled up onto the porch just as the metallic scrape of the cage bolt sounded.

The men around the circle all began hooting and hollering, shoving and moving in one body. The men in the back pushed to get in closer, while those with the best view suddenly found themselves in proximity to a starving, maddened bruin, and desperately tried to claw their way to the back.

Jack stood steadfast in the middle of it all, a modern day Beowulf awaiting the clash with black Grendel.

The bear batted aside the cage door and lurched out, its nose twitching in the air. A line of saliva trailed from its black lips to the street. When I'd first seen it, it had been a healthy specimen. Now, starving in the mud, exposed to the elements, the top of its head was balding from mange. Its eyes were cracked and bloodshot.

Its long neck swung back and forth as it sniffed at the crowd and growled, its teeth popping now and again from sheer nervous anger.

From my vantage I noticed the naked scar that ran down the bear's shoulder. It was an old wound, long healed over.

I remembered Jack's words as he hunched down and began making deep baritone barks to lure the bear's attention.

"Cain't ye see? Thet's my bar..."

Then the bear saw Jack. For a moment its jaw, which had been open and roaring to blot out the din of the cackling men, closed. Its tongue slipped out docilely for a moment and smacked across its lips. It gave a dip of its snout and a short snuffle, and went straight at Jack.

Jack met the bear with wide open arms. The animal surged up on its hind legs. Jack pitched his head down beneath the bear's chin, and the bear wrapped its powerful forearms around Jack's shoulders. The hooked tips of the inch and a half long claws disappeared in his flesh. I heard his muffled scream, and saw his muscles bunch and strain.

Jack's knife hacked into the hide of the bear. Red stripes ran in rows up the Missourian's back and the bear howled with fury as the bent blade struck home and returned gory, only to fall again.

The lusty crowd threatened to drown out the reciprocal screams of man and bear. Blood, skin, and fur alike were torn as the two massive creatures battled, locked in a vulgar parody of a loving embrace. This was contention at its most brutal, with nothing to separate man from beast but one simple tool that was not near as practical as the natural razors of the bear's dreadful claws.

The bear's blood drenched Jack's right hand, streaming from the great wound that his blade had worked open in its side. The agonized growls of man and bear ran together, as did their blood.

It seemed Jack would be sliced to ribbons. The bear kept swatting at his head, glancing him with its huge paw. The blood was running in his eyes now. Jack struggled to keep the bear on its feet. He planted his foot on one of its back paws, to keep the animal from using its full weight to bring him down. If he lost his feet, he would surely die. One arm strained to keep the bear's arms up, for if those keen claws found his belly, they would spill his guts sure as a toddler will spill his milk. His blade kept excavating.

Perhaps War Bag's insanity had become contagious, for I began to understand with a dreadful conviction that what Jack had told me was true. This *was* his bear. They *were* childhood enemies. Both had crossed miles in distance and circumstance to meet together one last time. That hairless scar on the bear's shoulder was where Jack had told us he had left his knife the last time they had met, somewhere amid the piney mountain cathedrals of his Ozark plateau.

It was all true. Everything unfathomable, everything impossible. It was all true.

From behind me, there was a womanly scream.

Cathy stood on the boardwalk, the horror plain on her face. Her palms curled against her cheeks, and her fingers were slatted across her wide, terrified eyes.

"Oh my God!" she screamed. "Somebody ought to stop this! Somebody ought to stop this!"

She was ignored.

I was sick to my stomach. I got down in the street and felt the ground around me. I found Jack's Winchester. But when I stepped back up on the porch and levered a round in, I knew I could not shoot without hitting Jack.

"Jesus *wept*!"

I turned. There stood Fuke, his eyes wide and fearful, his teeth clenched beneath his mustache.

I thrust the rifle at him.

"Stop it, Fuke!" I begged.

Fuke took the rifle without hesitation. He flipped it up against his cheek and took aim.

As though Providence had decreed it at that moment, Jack and the bear pushed away from each other.

Fuke fired over the heads of the crowd. He didn't wait for a reaction, but levered the Winchester and cracked out another shot.

The crowd fell flat. At first I thought he had blown off the top of someone's head, but then I saw a man grabbing for his hat. Fuke fired three more shots, as quick as thought.

When the smoke cleared, the bear was lying on its side, its tongue lolling from its mouth and blood spreading out in the street.

Jack fell to one knee in front of the bear, and thrust his bent knife into the ground to steady himself. His naked shoulders were painted with so much red it looked as though he'd been skinned alive.

Fuke handed the rifle back to me.

The men in the street all began to sit up, looking cautiously back at the porch.

Cathy buried her face in Fuke's shoulder.

Somewhere, a lone pair of hands began applauding.

Every head swivelled to find the origin of that audacious sound. Leaning against a tree across the street was Jinglebob Beddoe.

He was very drunk, the brown glass bottle of whatever he had imbibed tucked under one arm. He had on his pistol, and there was a crooked smile rippling along his milky face.

"Real good shootin,'" said Jinglebob to Fuke. "Real good."

Fuke's eyes slid to look at me.

Just then I saw Roam loping up the street on Crawfish.

Jinglebob began to walk deliberately across the street, teetering from drunkenness.

"Kinya hit the mahk everytoime, mate?" he asked. His accent, subdued before, was barely comprehensible.

Fuke looked at Jinglebob apprehensively.

Jinglebob took the bottle in his hand.

"Kinya heet this bottle?" He motioned to me. "Give eem the roifle beck."

I looked at Fuke.

"Dew it!"

I held the Winchester out to Fuke.

Fuke took it, but made no move to raise it.

"Big Indian foightah," Jinglebob smiled. He flung the bottle across the street then, and it smashed against the wall, showering Fuke, Cathy and I with broken, booze-smelling glass.

Cathy screeched and ducked into the Bee Hive.

I backed away, rubbing at my cheek, where a shard of glass had striped it.

The men scurried away like rats exposed, taking cover behind barrels and the corners of buildings. Only Jack remained on all fours, bleeding.

Roam approached steadily. He did not mark what was going on.

Jinglebob stepped around Jack and came to stand in the street about ten feet from the porch.

Fuke clutched the rifle tightly.

"Oi had twenny dollas 'said thet beah was gonna teah a hole in that dawg-killin' bastad."

He reached into his pocket and produced a brown wallet. He waved it at Fuke.

"Now Oi'll bet the same twenny roight heah. That Oi can blow you into the next world, fancy man."

Roam drew to a stop when he saw Jinglebob and Fuke facing off. He looked at them both anxiously, his eyes darting back and forth in his dark face.

"Fuke?" Roam said.

Fuke just stared down at Jinglebob. His lips were trembling.

"Fuke!" Roam said again.

"Keep outta this y'bloddy wog!" Jinglebob cursed, not bothering to look at Roam. He nodded to Fuke. "Cam on, yeh tin arse! Fill yor hands!"

Jinglebob's hand jerked out his pistol, and a shot echoed down Griffin Avenue.

The next instant Jinglebob lay on his side in the street, his legs crossed over one another like a ballet dancer caught in mid-pirouette. One arm was bent over his ragged ear, the pistol in his open palm, his slouch hat crushed beneath his head. One eye stared at the ground and did not blink. The other was screwed up to heaven and spattered with mud. There was a red spot growing in his left side, under his armpit. He had been blown clear out of his boots, and his bare feet were pale and clean as dove wings.

Fuke stood on the porch, Jack's rifle still lowered. He hadn't moved. He was staring at the dead man.

I felt my back touch the support pole of the Bee Hive's porch.

The street was mortally still.

Roam stood in his saddle, his pistol smoking in his hand. While I watched, he lowered it slowly.

Poking from behind the rain barrels and around the corners of the frame buildings and from behind the wagons where they had all taken shelter, the men of Griffin appeared two at a time. Some men looked from Jinglebob to Fuke. Others looked at Roam, and these began to murmur.

I went to Jack, who had been kneeling there bleeding.

"Jack, are you alright?"

Jack looked at me, his eyes unfocused behind the veil of blood.

"Ah could've kilt him, Stretch," he said.

"Roam!" I called, "Help me get him to his feet!"

I got one of Jack's bloody arms over my shoulder and hoisted him up. He clutched the bent knife tightly and kept looking at the dead bear.

"Roam! Fuke! Help me!"

Fuke stood on the porch.

Roam sat on his horse, watching me.

The men stood flanking the street on the boardwalks, talking to one another and gesturing to Roam.

I looked about, beginning to guess what was about to happen.

"Fuke!" I called. "Say something!"

Fuke sat down on the porch, rubbing his forehead with his hand. His face was drawn. He looked sick.

"Say something, you son of a bitch," I growled. "He just saved your life!"

Then it began.

"Lynch that nigger!"

The men poured from the sidewalks out into the street in one body, their faces tight with anger.

As they swarmed around him, Roam's face fell, and his eyes were like those of a tired parent thronged by a troop of haranguing children.

I saw someone reach up and jerk Roam down from the saddle. Before he disappeared, another hand took his pistol.

Someone snatched his Spencer from its scabbard.

Crawfish reared and crow hopped, panicked by the sudden rush of bodies.

Someone grabbed his bridle and forced his head down.

I stared at Fuke. He put his head in his hands.

I saw Roam's fists swinging, and his contorted face vanished. The mob flowed toward the river, bearing him along. Crawfish's painted coat flickered in the swell of clenching white hands, angry faces and bobbing hats.

I hobbled along behind. Jack leaned on me in a daze, stumbling and bleeding over my clothes.

Fuke stayed where he was.

As we hurried to catch up with he mob, I saw people craning their necks to watch. The Tonkawa boy scurried past us, a coil of hemp rope in his hand. He was laughing, sure he could earn the favor of the fickle white men again.

"Come on, Jack!" I hissed. He gritted his teeth and began to rediscover his legs.

The mob surged down Griffin Avenue, the shops dwindling away. The river crossing's grove of tall trees loomed like waiting executioners, sure doom for Roam, who was buffeted along on the tide of shouting men like a dark twig on a white current.

The road spilled out into the crossing, little more than a dusty bank. The mob halted short of the shimmering water.

I saw someone take the rope from the Tonk boy and it too was gone, hungrily swallowed by the many armed creature that swelled and kicked before us like some amorphous, indistinctly human amalgamation. It took seconds for a practiced hand within to sling a noose over the bare limb of an old tree.

The loop swayed as the immense human monster before me sucked it taut. It hung there, braced to pull the life from a man and render him a revolting, artificial blossom (a *cottonwood blossom*, I realized, numbly), no fruit but for the scavengers.

Crawfish was passed up the line and positioned under the tree. Then Roam reappeared, ascending above the crowd of men, bloodied and beaten. He was returned to his saddle, his arms bound to his side with a thick leather belt. His own gun belt, I saw.

I let go of Jack and heard him tumble to the ground as I bolted forward, drawing my pistol and firing into the air.

All heads turned to me, and for a moment I thought I would stop them. Then four men at the back rushed out and tackled me to the ground. A knee drew up into my stomach and all strength left me.

"Nigger-lovin' Yankee *shit*," someone whispered in my ear, and followed it up with a hawk of saliva that splashed into my ear.

The shouts reached a crescendo. I struggled to see Roam. His head was down, but his eyes glared from under his brow, unblinking and defiant. Blood trickled from his nose and his lips, and a swelling on his cheek bespoke the impending birth of a bruise that would never heal.

The rope drew tight, forcing him to sit up. I saw a hand in the crowd raise a pistol as the others cleared a path in front of the wild-eyed Crawfish, who was ready to bolt.

The thumb drew back the hammer and I was pushed back down, I heard the word 'justice' uttered.

Fifth

With Finality Came The Thunderous Report of a gun. But it was no revolver. It roared as though God Himself had taken exception. The thick branch above Roam's head blew apart in a spray of splinters and frayed rope and dangled from his shoulder.

Crawfish screamed in terror as wood chips showered down all around him, legs kicking out and sending the would-be executioners flailing away. Roam's hands clenched the saddle horn and his knees locked, and it was all he could do to keep from being thrown. He cursed and drove his heels into the pinto's sides. Crawfish took off like a skyrocket right past where I lay under a blanket of irate men.

The man who had spit in my ear sat up, releasing my pistol arm. I swung out, knocking the hot barrel of my Volcanic across his temple.

Seeing the gun, the other two on me drew back in fear like wolves before a fiery brand.

I sat up and waved the pistol at each of them. No more bullets would I sling like a fool at angels in the clear sky; not when there were so many devils here on earth who deserved a blue pill to calm them.

I got to my feet, and saw War Bag, with his hard-faced old Tonks, and the intense young Trigueño. The three of them were mounted, and the old Tonks had their cap and ball pistols drawn and cocked. War Bag stood, the barrel of Bullthrower beclouding the air with acrid smoke. He had fired over Solomon's saddle, and the obedient, brave mount had not stirred.

Like a dutiful squire, Trigueño pulled the old man's Henry rifle from the saddle scabbard and handed it down to him, hefting Bullthrower in exchange.

War Bag jacked a cartridge and aimed at no one in particular. In his eye I saw that he was not intent on shooting any men today, but that he would take no small delight in wounding that great multi-legged beast that had almost claimed the life of his friend.

I stepped back and joined them, adding my Volcanic to their arsenal.

War Bag gestured to Trigueño and the boy cut away from us, galloping after Roam. Crawfish had lit up the riverbank, and for a few tense moments we stood facing down the glowering eyes of the beast while Trigueño caught hold of Crawfish and mollified him.

No one spoke. A chill wind, pregnant with the scent of rain, slithered through the clacking tree branches. In the eyes of the men I saw impotent frustration. Not one of them would offer challenge.

Trigueño returned with Roam loping alongside.

The black man took sight of War Bag, and dipped his chin once. It was all the thanks he gave, and more than the old man asked.

Jack stumbled up on his uncertain legs, still awash in blood and bare chested. I hugged him, steadying him on his hide bound feet.

"I'God, Fatty," War Bag muttered. "What in hell happened to you?"

Jack's watery eyes shined at the old man, but he did not speak.

War Bag handed me his rifle, and pulled himself up onto Solomon's back. He reached out his hand and Trigueño gave him Bullthrower.

I noticed two pack horses being led by the Tonks. Trigueño's lance was tied to the saddle of one. Without a word I led Jack to the other and he crawled wearily atop it. I handed him the Henry, stowed my pistol, and climbed on top of the other horse. The Tonks tossed us the tethers. Jack leaned far forward in the saddle, but took up the reins.

In a commanding voice, War Bag said,

"Any one or ten of you that tries to follow us . . . . won't get far."

Then he turned Solomon around and headed back up Griffin Avenue. Trigueño filed in behind, then the Tonks, then Jack. I shared a look with Roam, and we fell in alongside each other and went with the group.

Neither man nor bullet followed us.

There was a rumble of thunder as we rode up the street. People on the sidewalks who had watched the first parade with mild interest, now stood agape at the sight of our's. We were a proud procession, contemptuous of all that we passed. I had never seen Roam's eyes so mean.

The first specks of rain were striking the street as we came within sight of the Bee Hive. Someone had cut off the bear's claws and dug out its teeth.

"They ought not to have done that," Jack remarked hoarsely.

Fuke still sat on the front stoop. He slowly got to his feet, his eyes incredulous.

War Bag did not slow his pace. He spared one look at Fuke as he passed.

Fuke's hands moved uselessly at his sides. He opened his mouth to speak, but War Bag was gone up the avenue.

Jack murmured,

"S'long, Fuke."

Roam did not look at him.

Fuke stepped off the porch and walked out into the center of the street. He was still there, a dot in the center of Griffin Avenue watching us, when we crossed Collins Creek. We steered northwest out of town, face first into a driving wet norther.

Soon we were far from Griffin. We stopped to tend to Jack. He was running a fever and shivering, though War Bag had wrapped him in a slicker and blankets from one the packs. The big hand that held the blankets closed was pale and wrinkled from the icy rain.

"I'm cold, Stretch," Jack complained in a tired voice.

"You'll be alright, Jack," I said dubiously.

We cleaned his cuts as best we could, and War Bag found a wool shirt that barely fit him.

We wrapped him in the slicker and the blankets again and pressed on. He kept slipping from the load, and I did my best to keep him aright as he had done for Fuke back on the Picket Wire.

We rode for some time with the two old Tonks in the lead. At times they would ride far out of sight, until I would worry that the Indian party that could any moment come over a rise and intercept us on its ride to Griffin had caught them. But they always returned.

It was high noon when one of the old Tonks (Yellow Band, as I had come to think of him, for the faded band of cloth that encircled his head) came riding up to War Bag and unleashed a flurry of excited gestures on him.

War Bag answered with some silent query, to which Yellow Band held up one finger.

War Bag turned to us, and signaled for us to dismount.

We led the horses quietly forward for about a half mile when we saw the other old Tonk, Great Coat (so-called by me for his outer wear) signaling to us up ahead from where he stood beside his pony.

The body of Dobie Flood was lying face down in the grass, pinned to the earth by arrows. His trousers had been yanked down unceremoniously around his ankles, and what the Indians had done to his privates I can hardly bear to recall. I noticed he had been stripped of his splendid Indian trinkets, and his feet were bare. The soles had been burned with heated points.

"Dobie," Roam said, squatting down beside the corpse and scratching his chin whiskers. There was a hint of regret in his voice.

"Could it have been Comanches?" I asked.

Roam ran one of the arrow shafts through his two fingers.

"Dog Soldiers." He sucked in his lower lip. "Where's yo tribe now, Doe?" he whispered.

Great Coat was kneeling in the grass touching the earth with splayed old fingers, thoughtfully perusing the ground as though it were a volume. He looked over at War Bag and began to sign.

"He says..." War Bag began.

"I know," Roam said, glancing briefly at what to me was only more prairie grass. "I seen it. Fourteen ponies headed northwest in single file. All of 'em unshod and carryin' riders." He looked at me. "'Cept one."

"That'll be Othello," I affirmed.

"Dobie said they was ten."

"Maybe they've got some Comanches with them?" I asked.

"Alright," War Bag said. "They've got a few hours ahead of us. We'll push on till dusk, then make camp."

He signed to the others, at the same time relaying his orders in Spanish to Trigueño. The Mexican boy nodded and was the first back on his horse, as though he were eager to see what lay ahead.

"War Bag," Roam said, standing up. "If they ain't headed towards Griffin, they ain't no use in trailin' em."

War Bag climbed atop Solomon and turned to fix his eye on Roam.

"Go on back there, then."

Roam put his hands on his hips and chewed his lips.

War Bag went on with Trigueño. The Tonks mounted up and followed.

"What do we do?" I asked Roam.

"You ought to take Jack on back to Griffin," Roam said.

"I'd never make it back," I said. I never felt my inadequacy more keenly than then.

"Jack could find it," Roam suggested.

I looked at Jack. He was leaning on the pack horse, staring up at the sky. His eyelids fluttered open and shut, his breath came in hot puffs from his slack jaw.

"Jack doesn't even know where he is."

"We got to rest soon. I think he hurt worse than he look."

"Ah kin make it till dusk," Jack said. The sound of his voice gave us both a start. He stared up at the sky. "Ah reckin'," he said, after a pause.

As though to demonstrate, he threw one leg over the horse, but could not summon the strength to mount. I had to help him, and almost tipped the struggling, protesting animal on its side.

Jack leaned forward, his forehead against the base of the horse's neck. Then he straightened, and grinned down at me.

"Awright. Let's go."

I got on the other pack horse and led Jack's. Roam fell in alongside me.

"I could guide you back," he said quietly. "Then turn around once we's in sight of town, or I could take you farther on. To Hidetown maybe."

"Where would you go?" I asked.

Roam blinked his eyes. He had killed a white man. He had no safe haven, now. Not in Texas.

"Mexico, maybe," Roam said.

I looked at War Bag.

"We can't just leave him," I said.

Roam opened his mouth to protest, then seemed to relive the events of that morning.

"No, *I* cain't, Stretch. But you could. Just turn yourself right around. Take Jack. You might run across a Army patrol, or some outfit bound for Griffin…"

"No," I said. "I'll stay with you."

"We both will," Jack muttered.

Roam nodded slightly, and we said nothing more to each other.

At dusk we halted to set up camp near a lone dead tree. A new moon rose vividly in the deepening sky, burning away the thick clouds and spreading a pale, eerie bloom all across the forlorn landscape. It rained lightly.

The Tonks took a canvas tent from the pack horses and proceeded to raise it. Trigueño stood by watching, while Roam and I helped Jack from the back of his horse.

"*Lo debemos que dejar,*" he said.

Roam glared at him.

"*Se va murir,*" he went on, gesturing to Jack with a careless nod, "*probablamente.*"

Roam ignored him, and he and I eased Jack down beside the tree.

Trigueño spat in the grass and unsaddled his horse.

Jack's mount was in a bad way thanks to his dead weight and the added burden of the packs. Its back was raw with saddle sores. It would not last another day's ride so encumbered.

War Bag decreed there would be no fire, and by his tone, no protest either.

A watch was set up, and Roam sat outside the door swaddled in a blanket with War Bag's hat on his head and his rifle cradled in his arms.

We crammed ourselves into the tent and shivered as dark came upon us, gnawing at pemmican and drinking creek water while the wind rippled the tent walls. It was dim and blue inside, and the rain pattered above us like spilling gravel. Jack's sleep was fitful.

The Tonks shared their food quietly. Trigueño did not partake. He took out his pistol, and set to inspecting the loads and powder.

"Where are we going, War Bag?" I asked, my voice sounding small and scared with the wind raving outside.

War Bag passed his canteen to Trigueño and said nothing. All their faces were silver in the moonlight that came through the tent door, where Roam sat outlined. Jack's breathing was harsh in the quiet.

I was not to be denied.

"I said, where are we going?"

War Bag's eye flashed at me as he chewed his pemmican. He took his pipe from his bag, striking a match that lit up his scarred face. The light was swallowed in the bowl and became a curl of smoke that filled the tent with a rich, earthen smell.

He puffed twice, then passed it to Yellow Band, who eagerly took it up.

"My uncle," War Bag said, his voice thunderous and naturally attuned to the howl of the wind all around, "used to tell me at length about the Crow and the Blackfoot. He said you could never learn enough about 'em."

Yellow Band passed War Bag's pipe to Great Coat, who held it up to each corner of the shelter before partaking.

"On revenge raids, they would draw blood from their calves with bone needles, to set themselves for the task ahead. The pain of their wounds would remind 'em, when they were ridin' hard through the night with empty bellies, just why they were doing it. That way there were no moments where the men would forget their anger and think of comfort. There couldn't be no comfort until it was done. Then they'd ride home with their scalps and the ponies they'd won, and they'd stop just within sight of home. They'd spend the night purifyin' themselves—washin' away the scent of death with burnin' sweetgrass and tobacco. In the morning, they'd rub ash on their faces, to show their people that the war fires in them had burned out."

Beside me, Jack moaned lowly in his sleep.

War Bag kept his eye fixed on me, and there was a power exuding from it that I could not deny.

My eyes slipped involuntarily to the two old, painted men. They smoked, unconcerned with the white words. Trigueño watched the old man, not understanding his words either, but feeling the meaning in his voice.

Great Coat passed the pipe to Trigueño, and he puffed it once and held it out to me.

I saw my own shaking hand reach out and take it.

Trigueño nodded, smoke wisping from his mouth like dragon's fire. I fitted the stem to my lips and watched the ashes in the bowl glow brightly as I took in the ardent smoke and felt it curl about my face, stinging my eyes.

I handed the pipe back to War Bag.

There was a look in his eye that reminded me of the day of our skinning match.

I exhaled, closing my eyes as the smell of the tobacco passed up my face.

"We'll pull out before sunup," War Bag said.

Sixth

The Flight Of An Arrow awoke me in the morning. The sound was like the hiss of a bird on wing, and was followed by the impact, hard, like the first probing strike of a woodpecker's beak.

Jack and I were alone in the tent.

Through the flap I saw Roam practicing with his bow on the old tree.

Jack was staring peacefully at the ceiling. His eyes saw me and he smiled benignly.

"Ain't slept so in a long spell," Jack said.

"You sound a lot better," I said.

"Ah dreamt o'home. Keep closin' my eyes...tryin' t'get the pitcher back."

I stretched and rubbed the sleep from my eyes. I stepped out of the tent.

Roam let another arrow loose. A fresh shaft quivered in the trunk of the tree. He lowered the bow as I stepped outside.

War Bag was dishing out a breakfast of cold beans to the Tonks, and Trigueño was saddling the horses.

The hide quiver was strapped to Roam's back now. He looked the part of a proud African hunter. I told him so, and he shook his head and burst out laughing.

"Sheeyit," he said, and went and got his arrows from the tree.

"Fats still asleep?" War Bag asked.

"No."

"If you two want breakfast, you're gonna have to eat it now."

I went back to the tent and found Jack still smiling faintly at the ceiling.

"Come on, Jack," I said.

Jack looked at me and shook his head.

"'Cain't ride, Stretch."

"Why not?"

"Ah jest cain't feel nothin.'"

I moved next to him.

"You're just stiff, is all."

"Naw," he said calmly. "Reckin thet 'ol bar done for me after all. If Fuke hadn't've shot him, maybe....just maybe Ah'd've got him too. Shoulda been me that done him in, not some stranger."

Tears filled my eyes. I had pinched his arm hard the whole time he spoke, and he had not reacted. The tips of his fingers were blue.

"We can ride back," I said, my voice cracking. "We can get you back to town. To the doctor."

"He's plum tired of sewin' me up, Ah reckin.'" His smile fell then, and he looked up at me. "Stretch, when ye get back, would ye try and find Whisper oncet more? Ah jest hate t'think of him goin' back t'the way he was."

From outside the tent, War Bag's voice boomed,

"What're you two sweethearts doin'?"

I started to answer, but Jack shook his head.

"In my pocket, thar's my tally-book. You take it?"

I pulled back his blanket. His shirt was soaked with blood. He had bled clean through his clothes.

"Aw, Jack," I said.

I reached into his pocket and found his Jew's harp, a whittling knife, and the tally book - a dirty little leather bound journal with 'Stillman Cruthers' embossed in chipped gold leaf on the front.

"Hey Fats!" War Bag called, his voice a little closer.

"Ye was the only one never called me outta my name," he said to me, his eyes still smiling while mine were spilling tears.

My shoulders shook and I tried to hide it from him with my hand.

War Bag's face loomed in the doorway, scowling. When he saw me bent over Jack, my hand tight on his arm, his face fell and his eye passed over the Missourian.

"Go on outside," he told me.

I looked again at Jack, at his massive, herculean body. It seemed impossible that he could not summon the strength to move. There was not even a hint of the death rattle in his voice. Yet he was so bloody...

Jack just kept on smiling, as though it were all a matter of course.

"G'wan, Strey-etch," he said. "S'awright. Be seein' ye."

I stumbled backwards, and War Bag caught me from falling.

"What're you going to...?"

"Get out, I said," said War Bag gruffly. He pushed me outside.

I fell through the flap on my back.

"What's the matter?" said Roam, coming over.

"Jack is..."

A gun cracked, startling the horses.

War Bag emerged from the tent, Jack's belt thrown over his shoulder. He had the Winchester in one hand, and Frenchy's pistol in the other. It was smoking.

"Get up," War Bag told me, his lips tight. "And see to your horse."

Roam and I watched in numb silence as War Bag walked past us and tossed Jack's belt with his bent knife to Trigueño. The boy caught it easily, turned it over in his hand, and buckled it on beneath his serape.

"Jesus Christ," I mumbled. "Jesus Christ . . . ."

Roam was staring at the tent. It was quiet and ominous looking now.

The two Tonks started to walk towards it, but War Bag shook his head, and motioned for them to mount.

"He wasn't a *horse,* for God's sake!" I cried.

War Bag said, "He was dyin,' Stretch."

He paused, chewing his lip awhile. "Nothin' we could've done but watched."

I put my head in my hands.

"Roam," War Bag said.

Jack's (Frenchy's) pistol sailed through the air and landed in the grass at Roam's feet.

Roam sank slowly to one knee and touched the pistol with his fingertips. He eased it into his hand, and looked it over thoughtfully, then cocked it.

He stood up.

I raised my head and looked at him. He was not pointing it, but everyone had heard the sound.

Trigueño's hand went under his serape, but War Bag touched his shoulder. Behind him, the Tonks on their horses swapped looks.

"What you takin' us to, ole man?" Roam said.

"Griffin's still the closest town," War Bag said. "I'll give you a share of food and water each, but not until we're done."

"Done doin' what?" Roam growled. "Huntin' Indians? You goin' get us all kilt!"

Yellow Band said something in his own language to get War Bag's attention.

He pointed back in the direction we'd come. There was a thin column of white smoke curling into the sky.

"Somebody's cookin' fire," War Bag said. "Couldn't be Indians. Even if they circled us in the night they wouldn't just tell us where they were."

He looked at Roam and I with his lone eye, thinking the same as we were – that Griffin had sent a posse out after us.

Roam watched the smoke. He eased the hammer down on the pistol and slipped it into his belt.

War Bag urged Solomon on. The Tonks took the pack horse that had been Jack's and Trigueño swung into his saddle.

I stood slowly up, unable to stop looking at the tent. It was a tomb of rough canvas that held my friend. He should not have drawn the end that he did.

I looked at Roam, and wiped the tears from my face with the back of my hand.

"Come on." He took me by the shoulder and led me to the pack horse.

We mounted up, watched the faraway smoke, and then turned our attentions to the tent.

"If it is Indians....," I said.

"It ain't Indians," Roam said. "But it ain't fittin' to leave him out here like this."

I took out the last of my makings and began rolling a cigarette.

"Anyone around will see," I said.

"Let 'em," Roam said.

I licked the paper and struck a match on my saddle. I took a long drag, and urged the horse forward. Roam fell in alongside me.

As we passed the tent, I tossed the cigarette into the open doorway.

Twenty minutes later we could see the smoke rising behind us in a voluminous pillar that blossomed into the firmament, a beacon for all but the blind.

War Bag turned in his saddle and glared angrily, but said nothing.

The day became unseasonably warm. The sun was high and blazing down. It dried out all our gear and the brown grass around us. We came across the bend of the Clear Fork and saw the first signs of a hunting outfit that had passed by only about a day or so ago.

Over the plain we saw a familiar sight; hundreds of dead and skinned buffalo, their humps cut inexpertly from their great bodies, leaving them ragged and torn. Some had been killed in the river, and their bloated carcasses left to putrify the water.

A few coyotes picked their way through the decimation. A dozen or more crows blanketed the huge, decapitated remains of what had in life been a proud bull, now no doubt a trophy being packed home in some greenhorn's buckboard.

An orphaned calf stood bawling near its mother, and Trigueño slid his lance out and chased it for a few feet before spearing it deftly through the neck.

We swam across the swollen river and the cold water was a relief. As soon as we reached the other side, the two old Tonks suddenly bolted forward on their horses, drawing their pistols.

The rest of our party watched in confusion. There ran a lone horse, chocolate in color, with a single half naked figure clinging to it.

War Bag slapped the reins against Solomon's flanks and the horse exploded into action, Trigueño's compact bay doing double time to keep up. The pack horse lagged in the rear, and I choked on the dust the others kicked up in their wake. Roam pulled ahead on Crawfish and for a while I was lost in a stifling cloud.

When my vision cleared, I saw that the chase had swung sharply back in my direction, only about twenty yards out. The quarry was a young Indian with a long train of wild black hair whipping behind him. He was the finest horseman I have ever seen to this day.

The Tonks, with reins in their teeth, began sending a clatter of pistol shots his way. He laid low against his mount's back and hung first from the left and then from the right, flattening himself like a tic, presenting no target.

On the climax of the second volley from the Tonks, the young Indian produced a bow and began launching flights back at his pursuers with amazing skill.

Great Coat's horse caught an arrow and collapsed, sending the old man flying. Yellow Band kept up, but soon his inferior mount grew winded and he gave the young brave distance.

Trigueño came speeding up on his bay like a mounted knight, his lance poised to strike. It became a contest of horseflesh then, and the pony of the Indian fed a steady diet of ground to the short legged bay, until it became clear that he would soon outrun the young Mexican too.

Then there was a shattering boom in the air, and the Indian flew from his horse. The horse kept running, and Trigueño galloped ten feet past the Indian before he drew to a stop and looked back in bewilderment.

To our left was War Bag, Bullthrower propped over the saddle of Solomon.

He waved for Trigueño to return, but the Mexican boy dismounted and knelt by the body of his rival. Jack's big bent knife flashed in the sun.

War Bag glowered, and when he saw us staring at him he shouted,

"This ain't no goddamn circus!"

In a few moments Trigueño came over. Tied to his lance was the Indian's scalp, and hanging from around the Mexican's neck was an impressive elk tooth necklace.

War Bag was sliding Bullthrower back into its case when Trigueño stopped before him. He lowered his lance to display his trophy, in the manner of a proud son sharing his accomplishments with his father.

War Bag glared up at the boy, and took hold of the lance, dragging Trigueño from his saddle.

"*Esto no es un juego!*" he swore into the boy's face.

War Bag took up the boy's lance, put the tapered point under his boot and using both hands to lever it, broke it in two with a splintering crack.

Trigueño stared quietly.

War Bag stood up and looked over at the Tonks. Yellow Band was standing over Great Coat, who was sitting in the grass clutching his knee.

"You goin' shoot him?" Roam asked dryly.

War Bag ignored him. He climbed upon Solomon and rode out to see the damage.

Trigueño got to his feet and picked up the two pieces of the lance. He felt them in his hands and then cast them disgustedly into the dry grass, as though he had decided War Bag was right. He got back on his horse.

"What do you suppose makes him stay?" I asked Roam, meaning Trigueño.

Roam shrugged.

"We got to get away, Stretch. Them Indians goin' to come lookin' for that young buck, and when they do . . . ."

War Bag rode up with the Tonks. Great Coat's leg was bound in shreds of cloth and a stiff, improvised splint, but he showed no sign of pain.

"He's busted his leg," War Bag said, as he rode up and dismounted.

"What we goin' do?" Roam asked.

"Lay across the horses," War Bag said, and made a fanning motion across the field of stinking buffalo. "Spread out as best you can. Their trail leads that way, so that's the way they'll come. I don't want 'em to see us till it's too late."

"What do you intend?" I asked.

"They're going to come back lookin' for that boy buck. When they do, we'll kill 'em all."

Seventh

An Uneasy Hour of waiting followed.

We tied sodden bandannas around our faces to keep out the stink of the dead buffalo. Roam planted his arrows in the dry earth and leaned his bow near to hand. He had Jack's Winchester now, and a box of cartridges which he spilled in his pocket.

The young Indian's pony was tethered to a spot in the field so it might bait the others.

Keeping the ponies still and lying in the sun for what could be a long wait was impractical. We hunted about and found a trickling stream at the bottom of a low gully about thirty yards to our right, where the horses could stand and be out of sight. War Bag did not like the idea of making our only means of escape or pursuit so vulnerable. Great Coat, because of his leg, was elected to stay with them.

Yellow Band finally agreed to remain with his partner in the gully.

"That old boy's itchin' to fight," Roam whispered to me.

Trigueño took War Bag's Henry rifle and lay behind a carcass in front of Roam. There was barely contained excitement in his eyes.

We were in position the better part of an hour when I saw a flash in the distance at ground level. I signaled Roam, but he had seen it already.

War Bag took out a small mirror and held it up. He tilted it briefly towards the sun. Bullthrower sat propped on a rest stick, as though we were waiting for a buff herd, and not a band of Indians.

War Bag put the mirror away and we watched. There was no reply.

I was greasy with sweat, and the swarm of flies from the rotting buffalo were thick as a black fog in the air.

After about a half an hour, a single Cheyenne came loping over, plain as day.

He was about the age of the one War Bag had killed, and seemed decked out in his best regalia. He sported a hide shirt painted with colorful symbols, and was ornamented with various feathers, furs, and beads.

The youth slowed his horse as he got close enough to be discernable. He threw up one hand across his eyes to shade them. The angle of the sun was to our advantage.

War Bag turned to look at Roam, and his expression was full of meaning.

Roam set down his rifle and picked up his bow. He fitted an arrow.

The Indian seemed uncertain. His face turned left and right, taking in the littered plain. Still his horse walked forward.

The brave smiled once, as though he thought it were a joke on the part of his dead friend. He was too young and foolish to be about such business. He got so close I could see a dried corn cob dangling from his saddle.

Roam drew back on his bowstring, and watched over the lip of the dead buffalo. Only his eyes could be seen above the red bandanna tied around his face. I could make out the beading sweat, and see the tendons in his arms standing out under his shining skin. His legs shifted slightly underneath him.

The Indian leaned forward in his saddle, peering past the empty horse. One hand started to pull the head of the mount around.

Turning, the brave provided Roam with a broadside target. Roam's legs sprung. At the same time he reached his full height, his fingers released and the bowstring twanged like a tuneless guitar.

The arrow was drawn from Roam's bow, as though the arrowhead were magnetically attracted to something in the jugular notch above the Indian's breastbone. The arrow knocked him into a backwards somersault off the rump of his horse.

The horse shook its head and stepped away from the body, but did not panic.

Roam stood staring at the body of the youth who lay face down in the grass with the point of the arrow sprouting from the nape of his neck like a poker spade. Dark blood bubbled from the wound.

I looked at War Bag. His brief smile was approving.

Roam sank back down to his knees, sullen.

"When this one don't come back, the rest'll come lookin,'" War Bag said. "We can pick 'em off."

"How about that posse behind us?" Roam said.

Then a shrill, distinct keening pierced the air. It was like the death cry of a diving eagle, and seemed to come from every direction.

"*Eyow!*" called a voice to my right.

It was immediately followed by a babel of answering cries and a salvo of gun fire like firecrackers popping off a string. Smoke wafted up from the gully to the right. I heard the horses scream in alarm.

Trigueño, who was closest, turned to War Bag and hollered.

"*Los caballos!*"

War Bag stood, intent on rushing to the two Indian ponies, who were looking about wild-eyed, startled by the explosive sounds.

Then a sharp pain as of a mosquito bite flared in my hip. Looking down, I was rewarded with the greatest shock of my career. There was an Indian arrow lodged in my left leg. I made a gasping noise, and watched in fascination as blood began to seep around the shaft.

Peering over the edge of a dead cow only ten or fifteen yards to my left was a man who looked to be part buffalo himself. Black horns curled up from the sides of his woolly head, but beneath that animal scalp was the bright white painted face of a man with black beads for eyes. A red pit opened as if by enchantment in his chalk white visage, and a sharp yelping like that of an excited dog's issued forth.

"*Oo! Oo!*"

I actually saw the second arrow speed towards me, spiraling, the sun shining through the black feathers on its shaft. I felt the breeze as it passed by my cheek, and may have imagined the brush of the feathers tickling, but I do not think so.

My mind processed the meaning of the bow clenched perpendicularly in the ruddy fist of the buffalo-man. One bronze arm curled up behind his head and an arrow sprouted between its forefinger and thumb. He fitted it in a smooth motion to the bow. The sun glanced off the bright tip.

I was aware of a sudden explosion of color and movement in my peripheral vision. Shouts and shots mingled in my ears. Bullthrower boomed.

I flung myself face down into the dry grass as the third arrow passed through the space of air I had only just evacuated. I wrenched my Volcanic from its holster and drew it up, shooting before I saw. The smoke curled, and when it cleared the buffalo-man was gone.

I tumbled over the animal carcass and collapsed on the other side, intent on putting it between me and the Indian who had flanked us.

Indians were poking their feathered heads up from the gully where our horses had been, and I saw Yellow Band crawling out, his body covered in quivering arrows. An Indian poised at the lip of the gully put a carbine to his shoulder and fired a bullet through the old Tonk. Yellow Band did not move again.

Then, like an Indian himself, Trigueño appeared from cover and leaped full upon Yellow Band's killer, his hands full of pistol and knife. The two fell back into the gully, out of sight. I saw Roam firing the Winchester, his bow cast by the wayside. The powder cloud from the muzzle formed a ghostly aura about him. Spent cartridges flipped end over end through the air.

The Indian ponies had been cut down. No - one was alive, and War Bag was laying across it to keep it down. It was the pony of the boy Roam had killed. Bullthrower lay useless in the thick of this fight. War Bag had a revolver in hand, and was flinging lead all around him.

Indians were flashing before my eyes everywhere I turned. They disappeared in the smoke and behind the buffalo carcasses, sprung up from the rear, fired arrows or rifles, and then seemed to meld with the very ground. The killing field with its grisly cover had become a disadvantage. The Cheyenne (and Comanche?) were utilizing it to greater effect. They knew the art of evasion and hit and run warfare. For the most part we were just panicked sportsmen whose game had unexpectedly turned on us, firing into the wilderness as it came alive all around.

They were forcing us to use our ammunition, and were spending their own sparingly. Twice I fired at Indians who did not shoot back.

I crawled on my face through the acrid fog until I reached Roam, the pain in my leg dull.

"We've got to stop shooting!" I shouted to him.

"They'll come in and get us!" Roam hollered back, pausing to fire at another ghost.

A bullet struck the horse War Bag was trying to keep calm, and it expired. War Bag cursed. With the Indians in possession of the mounts we'd left in the gully, we were now immobilized.

Then there was the unmistakable report of a big fifty from the direction of the river. The Indians heard it too, for one stopped and cocked his head at the sound, and Roam promptly shot him down.

There was another shot from behind us, this time of a carbine or other smaller gun. Again. Again. I craned my neck as much as I dared to see.

"Runners?" I wondered aloud. Maybe whoever had been killing these buffalo?

"The posse," Roam mumbled. "Gotta be."

And yet to me, it didn't sound like more than one rifle.

While the Indians paused to consider the intrusion of a second enemy, there was a yell of savage triumph from the direction of the gully. We turned our attention to it briefly, and from the ditch crawled Trigueño. He was cut in a few places, but there was a grin on his cruel face, and he scuttled on his belly like a crab to where War Bag lay.

War Bag whispered something to him in Spanish I could not make out. We all sat huddled, our nerves at end.

My heart beat soundly. My leg pained me, and the pistol was warm in my hand. I was alive. Though abhorred by what had led me here, and fearful of what might yet come, in my heart I found that I was glad to be there beside Roam and even War Bag.

Then with a start I saw the buffalo-man briefly appear, hurrying between the hoof of a dead buffalo and a dip in the land. I raised my pistol to shoot, but Roam, who had seen him too, touched my hand.

"They goin.'"

A grin played across my face and I looked at Roam. The sight of my expression proved infectious, and he shook his head and smiled.

"What the hell you got to smile about, Stretch?" he chortled, looking down at my leg. "You awright?"

"Yeah. I'm fine," I laughed.

War Bag slowly got to his feet with the help of Trigueño. I saw that he too had been wounded, by a bullet that seemed to have torn the leg of his trousers and cut across the thigh. He limped, but could stand. The top of Trigueño's ear had been cropped by a bullet, and the small finger of his right hand cut away.

Roam stood up. He had come through unhurt, save for a slight cut above his cheek which oozed blood.

The Tonks were not so lucky. Yellow Band lay where he had fallen, and Great Coat did not emerge from the gully. Our horses were gone as well. There was no sign of the Indians.

"Hope they brought spare horses," I said aloud, as I turned and watched the distant slope behind us where our saving fire had come from.

Then I realized who our saviors must be, and turned to Roam.

"You've got to get out of here."

"I got no hoss," Roam said.

"Then hide," I said. "We'll say you were dragged off."

"Ain't goin' hide from the likes of them," Roam growled.

War Bag limped up to us, leaning on Trigueño.

"Ain't gonna have to. Look who it is," War Bag said.

Over the plain, weaving between the stripped obstacles of dead animals, rode not a posse, but a single horseman. Fuke.

His capote was tied to his saddle, his Remington rifle hung in its scabbard, and his Winchester was bare across his lap. I did not recognize the horse he rode, or the meek expression on his face. His hair was in disarray, and the shadow of doubt was on his face as he came at last to within speaking distance. He halted his horse.

War Bag stared hard at Fuke, then nodded.

"George," was all he said.

Roam watched Fuke darkly and had no words.

I didn't know what to say either. As I looked at Fuke I still felt angry, and it didn't mix well with my relief.

"They ain't *following* you," Fuke said, his face earnest as a repentant child's. His blue eyes touched on each of us, desperate for a friendly look. "I told them you were headed for *Albany*. I swiped this horse from Yorkie Collinson." He snickered, but no one smiled.

Roam turned to me.

"You set down. I'll tend your leg."

I held onto my leg with both hands, still looking up at Fuke.

Fuke watched Roam for a moment, then nodded to War Bag. The old man leaned on Trigueño and watched Fuke quietly.

"It ah . . . looks like you found your *Indians,* old man," Fuke said. "Or they found *you."* He half turned in his saddle and motioned back over his shoulder. "I found a burning tent back *there,* but there was only . . . ." He looked around and tipped his hat. "Why, it was *Fats*, wasn't it?"

"It's Black Face, George," War Bag said. He hobbled closer with the help of the Mexican boy.

Fuke looked doubtfully at the old man.

Roam looked at me, and laid his palm on my thigh. He grabbed the arrow with his other hand.

I nodded.

Roam yanked the arrow out. There was a short raspberry eruption which he stymied with a piece of damp cloth. The pain did not subside, but it was good and hot.

"It isn't poison is it?" I whispered.

Roam lifted the hole in my jeans with his finger and let light on the bleeding arrow wound.

"Naw," he said. "Hold this tight."

I sucked air sharply through my trembling lips, as though it would cool the pain.

Roam stood and looked up at Fuke, who was still on his horse.

"You gimme that hoss," he said.

Normally such a demand from Roam would have brought a quick and bitter riposte to the Louisianan's lips, but Fuke remained calm, perhaps in light of the unsure ground on which he stood with us.

"Roam..." Fuke started.

Roam took out his pistol.

"I'll make it easy on you, swell."

Fuke bristled, but slowly eased himself from the saddle. He still held his rifle, but made no move to use it.

"What d'you intend, Trooper?" War Bag rumbled behind him.

"I'll be takin' Stretch south," Roam said, putting away his pistol.

"How about all that money in the poor box?" War Bag asked.

Roam moved to the horse.

"You do what you like," he muttered, and stuck his boot in the stirrup.

War Bag looked at Roam without a hint of surprise. His gnarled hand on Trigueño's shoulder clenched briefly.

Through a mouth of teeth, he said,

"Trigueño and me are goin' on."

"Goin' on where?" Roam asked. He had a hand on the saddlehorn and was set to pull himself up.

War Bag gestured ahead, in the direction the Indians had fled. There was a white patch on the horizon, as though the Creator hadn't yet gotten around to that part of the canvas.

"On foot?" Fuke exclaimed.

War Bag nodded.

"Right now they're regroupin.' Won't be long 'fore they turn around and come back to finish us. They know we got no horses. But they won't figure on us comin' to them."

"Goddammit, Eph!" Roam hissed. He leaned his forehead against the side of the horse.

"You all go on," War Bag spat. "But Trigueño and me, we're goin.'"

He turned around, the boy obediently keeping in step, and began to walk away from us.

Partly to my own astonishment, I called out,

"Wait, War Bag. I'm coming too."

Eighth

The Old Man's Zeal had somehow taken hold of me. I cannot hope to explain it in terms the reader will understand. Moments in our shared experience had chained me to War Bag's oars, and kept me fast even as he steered us headlong into the mouth of disaster. For as I watched him and the blood-soaked Mexican boy hitching purposefully toward that far-off whiteness, I knew that it was to oblivion he intended to go.

I remember to this day how he stood stubbornly against the tide of buffalo as Jack and I clung desperately to his heels while the sod trembled and the bulls snorted, and the hot brass from the Spencer rained on our backs. I remember the sound of his voice as he commanded the mob at Griffin, and a myriad other broken images that revolve like a system around the gray, fading picture of his face in my memory. His laugh at the campfire. The smell of his pipe. His booming voice in song. Lord, his *voice*.

Though there were times that his hold on me flagged, it never fully let up. Not when we'd learned of his abandoned family, not when the illusory fervor of his certain deathbed had been upon him, and surely not when he had dragged himself up by a cable of hate from the edge of that abyss. When he blew apart the hangman's tree, I knew in my soul that I would stand by him (and Roam too) through whatever end he saw fit to commit to.

I might speculate forever on how War Bag was like an old bull leaving the herd to meet his finish. Perhaps that promise of death was what he had paid the old Tonk hunters with, for they got no recompense for their trouble except a quick and violent demise. Maybe it was what he owed to Trigueño, whom he had taken from the Apache so many years ago, and who still bore the burden of that displacement. Perhaps death was what War Bag owed all of us. It was what he owed me.

I had come to live, to see and to taste life honest and full as it was to him. He had shrugged at my arrival like a coarse patron making room at the bar and begrudgingly allowed me to partake of his world, and now it was time to pay admission. There was more to this life. That is what I had learned in my odyssey. There was death. I would not have gone with him if I had thought it was mine; I honestly did not. But I found I wasn't afraid of it if that was the case. Whatever came next,

I had to see the end War Bag had planned for himself. If I could share in his death as I had briefly shared in his life, I would be . . . what? Whatever I had come to Texas to be. Shaped and fired by the old man's hand.

I stood up painfully while War Bag and Trigueño regarded me. Roam and Fuke were dumbstruck.

I limped over, the arrow wound pumping fresh blood down my leg in spite of the tight bandage.

As I took my place alongside War Bag, he nodded.

"You crazy too?" Roam exclaimed, staring at me. His foot was still poised in the stirrup.

"I don't know. I'm going with him, Roam. You can ride off, if you want to. I'm going with him."

Fuke lowered his eyes and snickered to himself, as though a joke he had never found funny before suddenly made sense. When he lifted his eyes, they were fixed on War Bag.

"Me too," he said. "If you'll have me, old man. I guess I didn't ride out here for *else*."

War Bag smiled thinly.

Fuke came to stand beside me, and we all fixed our attention on Roam.

Roam tucked in his lower lip. His eyes flashed on every one of us, denouncing us all as fools and yet pleading with the eyes of a child not to be left behind. It took a long time for him to decide, maybe because he had the most sense among us all.

But finally he brought his foot down. He led the horse over and tossed the reins to Fuke.

"You ain't goin' get far on that leg, Stretch," he said finally. "Goin' wind up on a peg."

Beyond belief, that was all it took. I say "beyond belief" because now I am a man of rational thinking, no longer moved by the irrationalities that sound from the passionate throats of young dreamers. Looking back I cannot now, with my rational mind, recognize who I was and why I played the part I played. But on that clear and unseasonably hot day I know that I was moved by a low, deep voice that spoke plain, even in what might have been the language of madness. We all did. That was all there was to it.

We walked, mainly in silence, and Fuke led his stolen horse behind.

We passed the canteens back and forth, and drank our fill, saving nothing for the return.

I thought of Jack and Monday and even Frenchy, how they should have been beside us, how in some way they were.

We made no attempt to hide. Where was there to hide in that empty land? We saw no tracks of our quarry. We would see none. We needed none. The buffalo led us. Dead, they were strung out in a long staggered line which we followed as sure as if they were great spoor.

I thought of the men hustling to plan their grand enterprises back in Griffin. Wright Mooar, Jim White, Collinson, and the rest. How would it end? As all hunts ended. As ours had. With the hunted strewn dead and stripped on the naked prairie.

And when there was nothing left to hunt, what would become of the hunters? Our stories would grow old with us, like fond friends. Indeed, we would treat them like old friends, as our mortal companions died off. And every year our stories would grow larger, and our audiences smaller.

Who would care to hear the stories of buffalo hunts when they could thrill to the midnight rides of the cowboy and the outlaw? And when they were gone, there would be some other. Who would remember War Bag Tyler, and Roam Welty? Who would recall Fat Jack McDade and his bear, and the jackrabbit in mid-leap that Fuke Latouche once picked off from the far bank of the Rule River? Who would laugh at the names of Monday Loman's mules, and marvel at his kindly mastery of them? Who would care who had not been there?

If there was anything less interesting than hearing the old stories of old men, it was reading their out of date words in musty books. What had I thought, all those months ago, when I left Chicago? That I would alight like a bumblebee on every 'character' I met, drawing forth the essence of his being and then returning to my little nest by the lake to expel it all into some fool book with which I would feed the hungry minds of readers I didn't even know, and thus achieve some love I would never need to requite?

My existence for once seemed truly real as I took that long walk amid the company of my friends. How clearly I can recall the sounds of our guns slapping against our thighs and the creak of our belts. The tickle of the wings of flies that buzzed past our cheeks was like the passing of fear-filled angels that fluttered about, begging us to turn back. Bullthrower pointed nonchalantly over the old man's shoulder like a warning to Heaven not to interfere.

I only heard one more bit of speech on the walk, and it was so low that I knew it was not meant for my ears.

"Roam," Fuke said. "Back there . . . I ever tell you about my *grandpa* who got killed by Sam Wilson at Cave in the Rock for chasin' a *nigra* woman?

"Yup," Roam answered. Fuke hadn't told him, but he had heard us that night.

"He *caught* her, Roam. *Understand?*"

All Roam said was,

"I know. S'awright."

I wondered briefly if War Bag had known. But of course he had. That very first night at Lyon, when he had led a stuttering Fuke out into the night, he had known. That was the reason Fuke was here now. Maybe he could betray Roam out of self loathing and envy and fear, but he couldn't turn his back on the old man. In the end, none of us could.

The brambles were like a field of bleached antlers.

It was the dead mesquite forest we had passed on the way to Griffin. War Bag had not lied about remembering the place. Even I had not remembered it until we were there.

The dry place before me, blighted like some dead, dark land, was a mocking reflection of my thirst, brought into being by wanting. I unscrewed my canteen and sucked the dregs.

It was dusk, and the mesquite took on a luminous quality. The sky was red and cast our vision in the appropriate light with which we were to see War Bag's final fury.

"Here's where I dumped the poor box," War Bag said, though none of us had asked about our lost pay since we'd set out from Griffin. "We split up, we ought to be able to find it . . . ."

His voice trailed off, his attention having been taken by something else.

It was not the few ragged buffalo that were standing there, though a moment later they became very important indeed. We had seen them amid the brambles, idly cropping the dead grass. They were old and worthless. About five of them, driven from the herd and thus spared the fate of their tribe. They were nothing compared to the phenomenon before us.

The red sun seemed to have touched the rim of the earth and ignited the dry prairie grass on the horizon, blowing a living wall of black smoke into the air. As we watched, it rose over the face of the dying sun in a sable wave. It sounded as if the gate to hell had opened and the damned were lamenting their torments all at once from the threshold. The grass was ablaze.

"Them crazy red bastards!" War Bag hollered, affronted that his enemy would have the gall to unleash such a force against him. He took off his hat and threw it down in frustration.

Fuke's horse began to shy, seeing the beginnings of the red fires as they lashed dancing into the sky, high and terrible.

The light of the rising prairie fire glinted in the eyes of us all. The far off roaring became like the fall of a burning rain, a sound like the Egyptians must have heard. As we stood there, a whole herd of elk came bounding past us through the mesquite. Birds took to the air and spread out, fighting to achieve the clear sky like drowning men seeking the surface. A snake slipped over my foot, and I noticed queer jumping movement in the grass all around me. Field mice. *Hundreds* of them.

The old buffalos turned their muzzles to the pungent breeze, snuffled once, and began to turn, lowing in alarm.

War Bag lifted Bullthrower to his shoulder, and I thought in his madness he was going to shoot uselessly at the fire. Instead, he picked off one of the buffalos. Its head rocked back and it collapsed. The others snorted and began to reel about.

"*Fuego!* Kill 'em all before they skin out!" War Bag roared.

All our weapons swung to bear on the old bulls. We blew a gust of lead into them, and they stumbled and bucked and shook their heads in the midst of it, their hides popping open where the bullets and balls bore into them.

Two died straight off, and the wounded bolted for elsewhere. Fuke dragged out his Remington and felled a straggler with a staggering shot.

Fuke's horse whinnied and bucked and jerked away from him, so that he had to drop his rifle and hold on with both hands.

"Let that fool horse, go!" War Bag commanded, drawing out his hip knife.

Fuke did not heed, but jerked the horse's head roughly down.

"He's alright!" Fuke shouted.

"Alright, hell! You better let him go or get on him, George! The rest of you cut into them buffs! Pull out the guts and crawl inside!"

War Bag fell to his knees and sunk his knife in the belly of one of the old bulls. He ripped it lengthwise with all his strength.

Roam ran to another bull and cursed, sticking the still quivering belly and doing the same.

Trigueño drew out Jack's big Oriental blade and began digging into the gut of the next closest bull.

I stood dumb, not sure of what to do. I had left my knife with the Indian I had killed and never bothered to replace it.

Fuke hollered my name. He had crawled onto the back of Yorkie Collinson's skittish horse, and offered his hand.

"Come on, *Juniper*! Get on!"

I shook my head.

"No! I'm staying!"

Fuke cursed and fought to control the pony. He looked at War Bag, who was up to his elbows in steaming buffalo innards. He looked at Roam.

"Get outta here, Fuke!" Roam yelled, pulling out a slippery mass of gut and shoveling it aside. "Get to the edge! Maybe you can pick 'em off! They goin' come at us behind the fire!"

Fuke nodded and yanked out his rifle. He slapped the butt of the horse with the barrel and was off like a shot, straight south, parallel to the advancing wall of flame.

"Roam! Gimme your knife!" I shouted, stumbling towards another bull I spied. It was one of the wounded ones, and had collapsed a little ways away.

Roam did not hear me. The conflagration was roaring now, like the rush of a waterfall. He was burrowing head first into the carcass like a grub.

I could feel the heat of the fire. War Bag's boot heels were disappearing into the rib cage of a bull.

Suddenly Trigueño was standing in front of me, and pushing his old knife into my hand. I did not bother to thank him, but took it and ran for the other buffalo while he went the opposite direction, back to the one he had gutted.

I slid to my knees in front of the buffalo and jammed the knife into the soft underside. Blood spurted, and suddenly I felt a hard impact against the side of my head that sent me sprawling, dazed. The buffalo still had some life in it, and had kicked me with one wild hoof. I was lucky to still have my senses, but not much. Soon I would be swallowed by fire.

I found the knife and crawled back to the buffalo, shaking the dizziness from me. It had sat up, and was struggling to find its legs. I slid through the grass like a serpent and struck it in the throat. The buffalo pitched dangerously and fell over again, taking me to the top as I held onto its beard, thrusting the knife again and again into the throat and the soft spot behind the ear. The buffalo groaned and shook. I had to grip its horn to keep from being gored or bucked away. It tried to rise again, but stumbled and fell, nearly on top of me. I could no longer hear its protestations. There was only the fire and the crackle of blackening grasses.

I do not know how long it took the bull to die. It still kicked feebly as I slipped the knife back into the belly wound I'd made and ripped far to the right until I lodged the blade in bone. I took hold of each rubbery, furred flap I had made and drew the stomach open like a pillowcase. The innards trembled in fear at the unexpected air and light, but I reached in with both hands, hooked into whatever was there, and drew it streaming, hot and stinking into my lap.

The weight of it all threatened to pin me to the ground. I reached in farther and pulled out more, scraping and tearing tissue to try and carve a space big enough to accept me. The smell was horrendous. The heat from the yawning cavity spewed full into my face, so that my vomit mixed spontaneously with the viscera.

I stuck one foot into the opening and paused to look over the carcass at where my comrades lay. The flames were almost upon them. I tore off my bandanna and tied it around my face, then ducked into the body of the buffalo.

It was warm and wet within, and slime filled my ears. I drew the flaps of hide together, but found they would not marry. Desperately, I began to rock the buffalo, hoping to put the opening beneath me. The legs wouldn't allow it. I turned my face inward, away from the light. I clawed at the gut that remained, forcing myself farther in until I was face first against the cavity wall, away from the opening like a burglar trying to escape detection in a nighted doorway. I was cramped, my shoulders up against the sides of my head. I could see nothing, and clenched my eyes tight. I coughed and took short breaths.

Then the heat came.

The slippery wall began to grow hot, and I knew the fire had reached me. My hands, which were pressed against the carcass wall, began to warm. My left heel seared, and I realized it must have been protruding. I jerked it in as far as it would go, wedging myself tighter. I could hear a large cracking sound, as of timber snapping outside. I tried to pull my sleeves over my hands. I bit at the cuffs through the handkerchief, wet with my own saliva, my only relief against the stifling lack of air.

I got the sleeves pulled halfway over my palms, and curled my hands into fists. My knuckles began to suffer, as though they were pressed against a steak that was still popping in the skillet.

I whimpered. The air grew stale and torrid, and I sucked in the handkerchief. I dared not cough and exert my suffering lungs. The edge of one ear began to scorch, and I was afraid my hair might catch fire. But there wasn't enough air for that, I thought. Not even enough to breathe.

My head was throbbing inside, eyes burning. My limbs began to jerk nervously, and I was afraid I would involuntarily jump out of the carcass into the fire. My elbows kept thrusting out, meeting sharply with the resistance of my airless buffalo-cell. My heart was beating a mad rhythm in my ears, and my brain was growing drunk on blood. If I passed out, I knew I would suffocate.

The air felt like hot poison to take in. My gasping lungs rejected it, and my stomach began to heave painfully. I swallowed bile and hugged myself tightly, as if I could keep my life from breaking out of its dying host. I had to have air now, before my limbs were too weak to extract myself. I did not know how much time had elapsed, or if the fire had passed. My ears were plugged with fluid, as though I were far underwater.

I began to try and turn back to the opening, but I had wedged myself in too deep. I tried to kick out with my legs, but my knees were lodged against some resisting protuberance. I was locked in a fetal ball, my neck bent and my chin jammed against my chest. I tried to work my whole body like a maggot trying to flee its putrid repast. I fought down panic with grinding teeth. My body shook. I tried to pull my arms back around, or push myself out. I forced them to touch the hot fleshy wall, and there was pain.

I pushed with all I had left. The world grew dim and dark and things seemed not to matter so much, when suddenly I felt my posterior pop out into the free air. My shoulders slid a measurable distance down, so that I could straighten out my neck. Gratefully, I worked my way out the belly and was born again flopping onto the black ground. The air, smoky and heavy as it was, was a blast of fresh spring wind to my starving lungs.

I lay my head down and sprawled flat on my back. My eyes were covered with a wet film. Then the sharp agony of burning licked over my right arm, and I knew I was on fire. I groaned and forced my eyes open. It was just a piece of burning mesquite. The fire had passed, leaving little islands of shrinking flame in a black sea of ashes. This little flame had been about to expire when I had thoughtfully donated my sleeve as fuel. I rolled quickly over it, and the wetness of my body smothered it.

I coughed. The air was overhung with a rich burnt smell and lingering smoke. All that had been white and pale was turned to black. What bushes of mesquite remained were tangled dark tendrils lit with trembling flames.

Then I heard rifle-fire, and I knew it was not over yet.

Last

While The Fire Was Over, The Fight Was Not.
I drew myself up to my elbows. The last thing I wanted to do was continue. Let the Indians kill me, I thought.

Then I saw, through my stinging eyes, a rider thundering across the distance, firing his rifle from the saddle. It was Fuke. He was charging in the wasted path the fire had eaten, levering his Winchester and shooting into a confused mass of what looked to be five Indian horsemen. The Indians were wheeling about and returning fire, slowly, as though they were unsure what to make of their lone assailant. He was riding in the old way, steering with his knees and shooting into them, the brim of his hat turned up. If only it had been Napoleon between his knees, and not that skittish nag of Yorkie Collinson's. Fuke and that horse had been like cloud and sky.

Two ponies galloped away from the skirmish riderless. One Indian on foot held the reins of a bunch of horses. He went down under Fuke's fire, and the horses bolted in all directions. I recognized the dappled pattern of Crawfish, and War Bag's black, Solomon, in the bunch.

I saw Roam and War Bag stooped over one of the blackened buffalo husks, busily extracting Trigueño from its stomach.

I found my aching legs and stumbled over, just as Trigueño emerged coughing, bloody, and covered in guts. The side of his face was badly burned along the jaw line, and the fire had eaten away the hair on one side of his scalp. Roam's right arm was burned black along the edge. War Bag alone was utterly unscathed, though you wouldn't have known it, covered in gore and blackened by soot as he was.

Trigueño coughed throatily and then lay still.

Roam looked up at me and grabbed my shoulder, asking me if I was alright. As he did, the skin flaked away from his forearm and tumbled down my sleeve.

I could hardly hear him, and dug with one bloody finger in my plugged ears.

"Fuke...?" I asked.

War Bag was watching the fight, his old hand closed around Trigueño's bandaged, lifeless paw.

"Cuss you, Lousiana," War Bag muttered, and his deep voice trembled.

The shooting had stopped, though smoke hung around the Indians like a fog. As I looked, I saw Fuke leaning far over in his saddle. One of the Indians, a tall warrior on a white horse, was trotting out to meet him.

I got my Volcanic out of its holster, it being all I had. Roam grabbed my gun arm.

"Don't! They think we dead."

True, the Indians were not looking our way.

"But they'll kill him," I whispered hoarsely.

Roam pushed my gun lightly down.

"He dead already," he said.

The tall Indian on the white pony reached Fuke. Before he could do anything, Fuke fell from the saddle and landed in a broken heap on the ground. The Indian leaned forward and looked down on him. He seemed to tap Fuke lightly with a lance. Then he took the reins of Fuke's horse without ritual and led it back to his band. They were only three now, rounding up the scattered horses.

War Bag bent over Trigueño and pulled back his singed, bloody serape, taking out the boy's pistol.

"What're you doin?" Roam said, staring at the old man.

"Look, they're leaving," I said, pointing.

The Indians were riding away *en masse* after the rest of the fleeing ponies.

War Bag checked the loads of the pistol, then stood up and raised it high in the air.

"I ain't dead yet, you bastard," he snarled, and cocked the hammer.

Roam jumped up and grabbed his arm.

"Ain't it enough? Just let 'em go!"

I stood up, feeling the pistol in my hand.

Roam had both hands on Trigueño's gun, and War Bag was fumbling to retain control of it. Their teeth gritted and their feet shifted in the ash. War Bag slapped his free hand on Roam's burnt arm and began to claw at the charred flesh.

For a moment more they locked together, the only sound their breath hissing out and their boots shuffling in the black earth. Then they parted. Roam fell backwards. War Bag raised his pistol up and pointed it at him. He fired.

Then I threw up my Volcanic and felt it hiccup in my hand, the sound ringing in my ear, and the flash burning onto my corneas. Even as I closed my eyes and smelled the smoke in my nostrils, I knew what I would find when I opened them again. My hand aiming the pistol shook.

War Bag sat on the carcass of Trigueño's buffalo, staring at me. I had blown a hole in his torso, just three or four inches to the left of his navel. The blood was pouring out of him, down the leg of his jeans. He drew his coat back, and stared at the mortal wound there. He was breathing through his mouth, and his face was as pale as his forked beard. He looked like any old man resting from a long walk, just for a moment.

Roam was standing, tense and ready, his hands held out and open in a warding gesture before him. Where War Bag's bullet went, if it hit him, if it was even meant to, I don't know.

The old man held Trigueño's pistol limply in one hand in his lap. He stared down at it as though he could not remember how it had gotten there. He looked over his shoulder.

My eyes followed his. The Indians had stopped and were conferring. They had heard the shots. How could they not? As we watched in silence, one broke off from the group. The tall one on the white pony. The other three remained in place, watching their companion. He rode casually through the smoking waste like a demon coming to herd a bunch of stray sinners back to Tartarus. His face was a black spot on his shoulders.

I looked at War Bag, and he looked at me. *Give me this*, his one eye seemed to ask.

I nodded, and backed away.

Roam's hands fell to his sides and he slowly sank to the blasted ground, sitting Indian-style in the ashes.

The tall Indian came over the smoking plain. His face was as black as his whipping hair, as though he had no features, only a long headsman's cowl of midnight cloth. He bore the lance he had counted coup on Fuke with. Maybe it was the same lance I had seen him use on Frenchy. His horse seemed impossibly white. It shined like a vision, or a living sculpture of mountain snow. Its mane and tail were as gray as rain.

War Bag pulled himself to his feet and staggered alone out to meet him. The blood fell in big droplets that sizzled faintly in the ash.

Soon Black Face was close enough to be real. Until then he had been a nightmarish figure to me, mostly fantasy, even though I had glimpsed him before. Now I saw the streaks in his hair, like cobweb strands floating in a river of oil. I heard how his cry, fearful though it was, was not as strident as the shrieks of the young braves he led. He raised his lance to his shoulder and gave heel.

The hooves of the white horse tore up the ground. War Bag's dying legs wavered at the onslaught of that charge, but his feet remained firm. He leveled the pistol and took aim with one hand, the other held to his bleeding body. Black Face bore down upon him, his teeth a bridge of white in his empty face as he called down death to take his enemy.

They met.

As the horse passed, it blocked my view of War Bag. I saw the plume of smoke spout into the air, and heard the shot ring out. The war cry of Black Face broke even as his strong arm came down like an axe.

When it was over, War Bag stood still, his head lolling on his chest, his face hidden by his hanging hair. Black Face's lance had transfixed him at a downward angle. The spear head was buried in the dark ground. It held his limp body aright, in a weird semblance of life. Blood ran down the makeshift stake. He slowly sank to his knees, down the bloody shaft, and died upright. His arms hung loose. The pistol dropped from his hand, smoking.

Black Face rode seven or eight feet more and then tumbled from the back of his horse. It galloped on without him. A cloud of ash settled on his body. The horse stopped and shook its gray mane, as if an evil cloud had left it. Blood speckled its bright shoulders. The breeze that had brought the fire down on us stirred the Indian's tangled mass of gray flecked hair like the caress of an understanding mother.

The remaining Indians turned without a word and rode away.

Then a shrill, feral wail sounded in my ears. I thought at first that Black Face had arisen, but it was Trigueño. He was on his feet and had the look of a crazed beast. He rushed at me, and raised in his hand was Jack's huge bent knife. The bloodlust was in his eyes. He had awakened and seen what I'd done to his patron.

I stumbled back, and held up my hand to ward him off. The knife whistled down and bit into my wrist, sending a lightning shock of unendurable pain streaming up the offended arm. I say unendurable because my mind, unaccustomed to such sheer agony, folded in upon itself and took my universe and all I perceived with it. I fell to the ground, and it seemed I fell even farther, for I saw and heard nothing more.

Epilogue

But I can't rightly say there was nothing more. Of course there was.

I remember more pain, and thirst and delirium, and I remember Roam there with me. He urged me along, carried me when he could, supported me when he could not. At some point he closed the wound Trigueño had given me with fire.

My clearest memory is of kneeling at a buffalo wallow with Roam and dipping my head down to partake of the filthy water through his bandanna. I remember the curly hairs swimming like amoeba in the gray water, and seeing pressed grass down below the surface at the muddy bottom. I remember the paisley pattern of Roam's red bandanna, and the sound of the water dripping down as it formed little circles. It was like the sound of coins dropping. It was buff tea.

Milt Sutton gave me the rest of the story, back in his cabin in Griffin.

Roam had killed Trigueño.

He and I had walked most of the way back. Or rather Roam had, for I couldn't have walked myself. He fought infection and the elements every step, until we met a patrol of Negro soldiers from the fort. I don't know if any of them knew Roam. I suppose someone must have, because he wasn't hauled back for the killing of Jinglebob Beddoe. Elijah brought me to Milt's cabin after the post doctor had treated me.

I didn't lose my leg, as Roam had predicted, but I still wear the scars of the buffalo-man's arrow, and the burn on my ear. My left hand was amputated, and thus, according to Orfeo, insured my own salvation. For it is a known fact among the *ciboleros* that the left handed don't get into Heaven.

For a good lot of weeks I mulled over the death of War Bag. Fuke and Jack and the others too, but mostly the old man. If he had not stood at last against Black Face he would have died from the wound I inflicted, and despite my rationalizations (I had saved Roam, I told myself, again and again), in my heart I knew it was murder.

War Bag was a rare breed, and I loved him dearly. But I knew we were all under his spell. If I hadn't broken it with a bullet from my Volcanic, Roam and I might have died out there with him. Like Trigueño, like Fuke and Jack.

This is what I tell myself.

I flipped through the tally book Jack left me, and found the journal Stillman Cruthers had kept, pencilled in before Jack's marks. For a week I read the life of a man I'd never met, and learned little except that he was much like I had been, and that he'd met War Bag when Mrs. Tyler had filed for divorce on grounds of desertion. Stillman's notes on the case, like all his entries, were brief:

September, 1860
Friday 6th

Met with Ephron Tyler today. Likable old codger. Did not have the heart to tell him his wife had moved on with a restauranteur. Case was short. He left her everything. Understand there is a son. C'est la vie.

The rest I knew.
I never saw or heard from Roam again. I imagine he made it out of Texas and to some safe haven, maybe with the poor box money, but I doubt it. I think the fire got our earnings that summer. Maybe he returned to the Indian Nation, and was there to see it turned over to the Union. Every year the Indian lands get smaller, it seems.

I became a writer of sorts, contributing to various newspapers from Texas to New Mexico, from Arizona to California. I covered the Apache Wars, and the removal of the Chiricahua to Florida in '86.

The life I lead now is a far cry from the one I'd imagined, and not near as satisfying as the one I led that summer, but it suits me fine. I'm married to a fine woman who overlooks my disfigurement and my distant tendencies, and I have a son who cares nothing for his father's long anecdotes, and would rather my grandson not encourage their telling.

The story of how I got to San Francisco isn't nearly as interesting as the story I have already told, but I suppose I ought to tie things up.

Milt Sutton told me flat out that it was unseemly that I should spend the rest of my days drinking and pining on an old nigger's floor, especially when that old nigger needed his sleep most nights. He had no patience for a white boy that stayed up into the wee hours squinting at books.

I went to Jacksboro and sought out Captain George Robson, a man who would be as much a mentor for my later life as War Bag and Roam had been till then. I got a job as a typesetter on his Frontier Echo, and impressed him enough with my book learning and verbosity to win an occasional byline.

I took one friend with me to Jacksboro. On my last day in the Flat, Whisper The Pisser came to me as meekly as he would to his mother's breast. I guess our mutual impairment must've forged some innate bond between us that hadn't been there before. So my promise to Jack was kept, if my promise to Monday was not.

That ragged old three-legged cat remained with me until 1883, the year and the very day I was to leave Texas. I found him curled up in a packing crate, sleeping the sleep of the ages on my old buckskin pistol belt and what souvenirs I had kept of my buffalo days. I sealed that box and buried it whole in the yard like a proper Egyptian.

In that time things changed in the Flat. A stone schoolhouse that doubled as the Mason's hall was erected, the red light district slowly faded out, and the land was settled. Griffin lost the county seat to Albany, but remained a trade hub for the hide hunters. They flooded the High Plains after MacKenzie destroyed the Comanche horse herd at Palo Duro Canyon.

The buffalo were utterly gone by 1880 or so. For myself, I never hunted buffalo, nor any other living thing, ever again. The remorse over the men and animals I killed that summer has never left me.

When the Captain made his move to Griffin in '79, we parted ways. Still, news of the Clear Fork country and the men I once knew always seems to find its way to me. I think that any word that triggers in me the memories of that time in my life is welcome. Even now, when the last of the buff runners are passing on, I still think of the prairie and the tide of buffalo that once moved across it, and the great men I knew who were swept beneath those waves and whose bones are there still.

As I wrote that last part, my grandson, who has often played his games in my study while I picked through the bloody and cracked old pages of my journal, brought me a fragment of a fossilized shell from off my shelf. It is dusty and broken. The other piece is long lost.

"Is this a fossil, Gramps?" he asked.

I told him it was, taking it and turning it over curiously.

"What beach is it from?"

It took me a while to recognize it, but holding it in my hand I finally remembered where it came from. It was the bit of fossil I picked from the bank of the Rule River in Colorado on the day we set out from Cutter Sharpes' for Ft. Lyon that spring. I told him it came from the Plains, and he shook his head.

"Sea shells come from the sea, Gramps."

"There used to be a sea there, boy," I tell him, using that sobriquet for young men which it is the obligation of old men to use.

"It's gone now."

—End